SpringerBriefs in Bioengineering

For further volumes:
http://www.springer.com/series/10280

Yu-Chen Hu

# Gene Therapy for Cartilage and Bone Tissue Engineering

Yu-Chen Hu
Department of Chemical Engineering
National Tsing Hua University
Taiwan, Republic of China

ISSN 2193-097X ISSN 2193-0988 (electronic)
ISBN 978-3-642-53922-0 ISBN 978-3-642-53923-7 (eBook)
DOI 10.1007/978-3-642-53923-7
Springer Heidelberg New York Dordrecht London

Library of Congress Control Number: 2013958452



Printed on acid-free paper

Springer is part of Springer Science+Business Media (www.springer.com)

# Preface

Tissue engineering and gene therapy are both perceived important milestones of scientific achievements over the past two decades and both disciplines have converged for over 10 years. The past decade has witnessed rapid progresses in the gene delivery-based tissue engineering, especially for the repair of cartilage and bone defects. However, to date there still exist roadblocks to the translation of scientific findings in the laboratory to the clinical setting, because of the concerns regarding the use of gene therapy vectors for the treatment of non-lethal diseases/disorders such as bone/cartilage defects. This book briefly summaries the current status of bone/cartilage tissue engineering, gene therapy concepts and vectors, and the combined use of tissue engineering/gene therapy for the treatment of bone and cartilage defects. This book also provides brief summaries regarding the hurdles for clinical applications and future perspectives. For this book, I would like to express my sincere gratitude to all the diligent laboratory members and my family for their full support.

Department of Chemical Engineering,
National Tsing Hua University, Hsinchu,
Taiwan, Republic of China

Yu-Chen Hu

# Contents

# Chapter 1
# Bone and Cartilage Tissue Engineering

**Abstract** Bone and cartilage are important components in the skeleton system, providing the major structure of the body of vertebrates and conferring protection and support of soft tissues. This chapter briefly reviews the constituents of bones and articular cartilages as well as cells associated with bone/cartilage healing. This chapter further introduces the concepts and critical elements of tissue engineering for the repair/regeneration of bone and cartilage.

## 1.1 Bone

### *1.1.1 Bone Components and Bone Formation*

Bone is distinguished from other tissues by the presence of inorganic crystalline mineral salts and calcium in the form of hydroxyapatite, and a broad range of organic components. The inorganic mineral part constitutes 65–70 % of the matrix and mainly accounts for the biomechanical properties, while the organic constituents comprise the remaining 25–30 % of the matrix [1]. The hydroxyapatite is initially laid down as unmineralized osteoid, and mineralization follows through the deposition of calcium and phosphate, which is catalyzed by alkaline phosphatase (ALP) secreted by osteoblasts. The organic constituent mainly consists of collagen I and other proteins including osteocalcin, osteonectin and bone sialoprotein, etc. Collagen I initiates and orientates the growth of carbonated apatite mineral, controlling its size and three-dimensional distribution (for review see [1]). The hierarchical geometrical structure of bone is critical for the mechanical properties and for cells which convert mechanical and architectural cues into intracellular signals, driving gene expression, protein production and general behavior [1].

Y.-C. Hu, *Gene Therapy for Cartilage and Bone Tissue Engineering*,
SpringerBriefs in Bioengineering, DOI 10.1007/978-3-642-53923-7_1,

Bone formation proceeds through either endochondral or intramembraneous ossification pathways depending on the types of bones. Endochondral ossification is the process by which mesenchymal stem cells (MSCs) differentiate toward chondrocytes (cells in the cartilage) and produce a cartilaginous template, which contributes to longitudinal growth of the majority of bones such as long bones. During endochondral ossification, chondrocytes proliferate, undergo hypertrophy and die. The deposited cartilage extracellular matrix (ECM) is invaded by blood vessels, osteoclasts, osteoblasts and bone marrow MSCs (BMSCs), followed by the initiation of mineralized ECM deposition.

Human long bones consist of two forms of bone tissues: cortical bone constitutes the outer wall to provide the major mechanical support and contains blood vessels, while cancellous bone consists of trabecular plates and bars that are found in the highly vascularized interior [2]. An adult long bone has a shaft (diaphysis) with two expanded ends and a large inner medullary cavity filled with bone marrow, fat tissue and blood vessels. In the diaphysis, almost the entire thickness of bone tissue is cortical and only a small amount of trabecular bone lines the inner surface of cortical bone and faces the medullary cavity [3]. Compared with mature bone (also known as lamellar bone), newly formed (i.e. woven) bone has a higher cell-to-matrix ratio and lacks structural organization. Both types of bone, however, are components of trabecular (spongy) and cortical (dense) bone. Human trabecular bone shows large spatial and temporal variations in structure and mechanical properties, whereas human cortical bone exhibits an osteonal architectural pattern and is stronger than trabecular bone under compression. The Young's modulus, a measure of a material's stiffness, is $\approx 17$ and 1 GPa for human cortical and trabecular bone, respectively (for review see [4]).

Conversely, intramembranous ossification occurs in the absence of a cartilage template and contributes to the formation of calvarial bone. The intramembranous bone formation initiates by the migration and aggregation of mesenchymal cells. As the process continues, the newly organized tissue at the presumptive bone site becomes more vascularized and the aggregated mesenchymal cells become larger and rounded, which is followed by the differentiation into osteoblasts. The osteoblasts secrete the bone matrix (osteoid) and become increasingly separated from one another as the matrix is produced. The newly formed bone matrix appears as small, irregularly shaped spicules and radiates out from where ossification begins as calcification proceeds (for review, see [1]). The entire region of calcified spicules becomes surrounded by compact mesenchymal cells that form the periosteum (the membrane that surrounds the bone). The cells on the inner surface of the periosteum also become osteoblasts and deposit osteoid matrix parallel to that of the existing spicules. In this manner, many layers of bone are formed [5].

### *1.1.2 Medical Need for Bone Repair and Current Treatments*

Approximately 6–6.5 million fractures are reported per year in the United States [6, 7]. Although bone fractures and tissue loss are able to self-repair [8], healing is problematic in more than 20 % fractures [9]. Approximately 10 % of all fractures and up to 50 % of open tibial fractures fail to reunite [10]. Additionally, large critical-size bone defect (>2 cm in humans) resulting from serious trauma or tumor surgery cannot spontaneously heal and union [8]. Massive traumatic bony defects have become increasingly common due to modern warfare and traumatic orthopedic injuries constitute the vast majority of injuries incurred on the battle field: 70 % involve the musculoskeletal system; 26 % of these are fractures and 82 % of fractures are open fractures [8]. To date, management of large segmental defects in the long bone still poses a tremendous challenge for orthopedic surgeons [11], partly because the injury impairs blood supply and results in ischemia, osteonecrosis, bone loss and ultimately non-union [2].

Currently available materials for bone reconstruction in the clinical setting include autologous bone grafts, allogeneic banked bone grafts and synthetic materials [4]. Autologous bone grafting is considered the gold standard for treating bone defects [9, 12, 13]. However, autografting is restricted by bone availability and the need for bone harvesting procedures. Autografting may cause donor site morbidity (e.g. infection, bleeding and chronic pain) [13] and the repair by autografting is not always satisfactory [14]. In a study wherein 30 patients have surgically induced long-bone segmental defects after tumor resection, treatment with vascularized fibular grafts produces primary union in 23 patients within a mean of 6 months [15]. However, more than 50 % of these patients have complications, and 40 % require re-operation due to non-union, graft fracture or infection [15].

Allografting can initiate a healing response and recruit cells from surrounding soft tissues to produce new bones at the host-graft interface [16]. However, allografting requires a contiguous vascular supply and adequate mechanical stability to allow vessel in-growth and eventual bone remodeling [17]. These conditions are often absent in traumatic defects where surrounding soft tissue disruption and instability are expected. Besides autografting and allografting, various synthetic bone-substitute materials, including β-tricalcium, hydroxyapatite, biphasic calcium phosphate, polymers and metals have been developed (for review, see [18, 19]). However, synthetic bone-substitute materials may result in poor integration, adverse reactions and eventual bone resorption [4].

Due to the limitations in each of these treatment options, bone fracture and massive bone defects still represent a significant cause of chronic morbidity, impacting individuals' mobility, health as well as social and economic status [1, 6, 16].

### *1.1.3 Bone Cell Types and Bone Healing*

Bone formation entails orchestrated cellular activities of osteoblasts, osteocytes and osteoclasts [20]. Osteoblasts derived from mesenchymal stem cells (MSCs) or progenitor cells from the adherent portion of bone marrow are responsible for synthesizing the organic ECM and regulating matrix mineralization. Osteocytes are mature bone cells accounting for over 90 % of adult bone cells [20], embedded within the osteoid [3] and function in mineral homeostasis, mechanical sensing and signaling.

Osteoclasts derived from mononuclear cells or macrophage [3] can resorb bones and play roles in skeletal growth and bone remodeling [4].

An important progenitor cell source contributing to bone formation is MSC. MSCs are multipotent stem cells capable of self-renewal and differentiation into different (e.g. adipogenic, chondrogenic and osteogenic) lineages under appropriate environmental cues. Upon commitment into the osteogenic lineage, MSCs differentiate into osteoblasts first and further differentiate to become osteocytes. The differentiation is accompanied by stage-specific gene expression pattern, matrix deposition, maturation and final mineralization. Alkaline phosphatase (ALP) is an osteoblast marker whose expression ascends early along the differentiation pathway but descends during the mineralization phase [21]. Osteopontin is expressed bimodally, with an early peak during the matrix secretion phase and another after initial mineralization. Osteocalcin is a late bone marker only secreted by osteoblasts and signals terminal osteoblast differentiation [21].

In general, healing of bone fractures involves (1) initial inflammation, (2) subsequent production of bone callus with poorly organized matrix for bony union, and (3) ensuing remodeling process that reshapes the bone tissues by removing, replacing and reorganizing cells and matrix (for review, see [2]). The initial week-long inflammatory phase of fracture healing is characterized by the infiltration of inflammatory cells including neutrophils, lymphocytes and macrophages, and the release of various cytokines and growth factors [1]. Inflammation in the early phase of fracture repair can contribute to healing by facilitating removal of necrotic tissue and by initiating repair, especially vascular invasion and cell migration. However, chronic inflammation has a deleterious effect on healing [1] and some cytokines (e.g. interferon $\gamma$ (IFN-$\gamma$) and transforming growth factor-$\alpha$ (TNF-$\alpha$)) have been shown to impair bone healing [22].

Repair involves the replacement of necrotic or damaged tissue by new cells and matrix, thanks to the activities of undifferentiated mesenchymal cells that migrate to the injury site. Soon after entering the site of clot formation and tissue damage, the mesenchymal cells proliferate and synthesize new matrix. Later they may differentiate into chondrocytes, osteoblasts or other cells. Repair of many acute injuries produces an excessive amount of cellular and vascular tissue with poorly organized matrix. Remodeling reshapes and reorganizes the repair tissue so that the newly formed woven bone remodels to lamellar bone. As remodeling progresses, cell density and vascularity decrease. The cells remove excessive matrix and the repair tissue matrix collagen fibrils become more highly oriented along the lines of stress. Most apparent remodeling of repair tissues ceases within months of injury.

One critical factor to successful bone healing is the formation of an extensive network of blood vessels [23], which is required for the transport of oxygen and nutrients to the highly metabolically active regenerating callus and serves as a route for inflammatory cells and progenitor cells to reach the injury site [24]. Therefore, angiogenesis, the growth of new capillary blood vessels from pre-existing host vasculature, plays crucial roles in the initiation of fracture healing and promotion of endochondral and intramembranous ossification in bone development/regeneration [1].

## 1.2 Articular Cartilage

### 1.2.1 Cartilage Types

Cartilage is a flexible connective tissue found in many areas in the bodies including the joints between bones, the rib cage, the ear, the nose, the bronchial tubes and the intervertebral discs. There are three types of cartilages: elastic cartilage, fibrocartilage and hyaline cartilage. Elastic cartilage is the cartilage present in the outer ear, Eustachian tube and epiglottis. Elastic cartilage contains elastic fiber networks and collagen fibers while the major protein component is elastin. Fibrocartilage is found in the pubic symphysis, the annulus fibrosus of intervertebral discs and knee meniscus. White fibrocartilage consists of a mixture of white fibrous tissue and cartilaginous tissue in various proportions. The major protein component in fibrocartilage is collagen I. Hyaline cartilage is the cartilage within the joints (articular cartilage) and is distinguished by the presence of a glassy, homogeneous, amorphous matrix. Articular cartilage is a durable weight-bearing tissue that provides frictionless motion between articulating surfaces while protecting the underlying bones from mechanical stresses [25]. Due to its importance in withstanding the mechanical load, the following discussions only deal with articular cartilage.

### 1.2.2 Cells and Components in Articular Cartilages

The only cell type in articular cartilage is chondrocyte, which is located within the spaces called lacunae throughout the articular cartilage. The articular chondrocytes/lacunae are embedded in the ECM comprised of collagens, proteoglycans, non-collagen proteins and water. Chondrocytes account for only 1–5 % of the total volume but are critical in synthesizing the ECM molecules such as collagen II (mainly collagen IIA1) and proteoglycans (for review, see [26]). The highly charged proteoglycans consist of different sulfated glycosaminoglycans (GAGs) that are linked to a core protein. Articular cartilage contains a variety of proteoglycans that are essential for normal function, including aggrecan, decorin, biglycan, and fibromodulin. The largest in size and the most abundant by weight is aggrecan, a proteoglycan that possesses more than 100 chondroitin sulfate and keratin sulfate chains. Aggrecan is characterized by its ability to interact with hyaluronan (HA) to form large proteoglycan aggregates via link proteins [27]. The proteoglycans keep the cartilage hydrated and impart the cartilage resistance to compression, whereas cross-linked collagen fibrils confer tensile strength to cartilage. The composition and structural arrangement of the ECM components confer cartilage its mechanical properties, and can vary with age [28].

Articular cartilage has a characteristic zonal structure which is classified as superficial, middle, deep and calcified cartilage zones, each having distinct cell morphology and matrix organization [29–31]. In the superficial zone, cells are densely distributed and oriented parallel to the articular surface, and appear more elongated than the cells in deeper layers. The proteoglycan content is lower while the collagen IIA1 content is higher than those in the deep zone. In the middle zone, the proteoglycan concentrations increase with depth while collagen IIA1 fibrils are less organized and aligned obliquely to the surface. The chondrocytes are rounded and more scarcely populated than those in the superficial zone. In the deep zone, spherical chondrocytes are arranged in a columnar orientation perpendicular to the joint surface. The proteoglycan content is the highest while the collagen IIA1 fibrils are less abundant and are oriented perpendicular to the surface. The tide mark distinguishes the deep zone from the calcified cartilage. The calcified layer plays an integral role in securing the cartilage to bone, by anchoring the collagen fibrils of the deep zone to subchondral bone. In this zone, the cell population is scarce and chondrocytes are hypertrophic [31].

### 1.2.3 Medical Need for Articular Cartilage Repair and Current Treatments

Degeneration of articular cartilage may occur due to trauma, metabolic or mechanical deficits which lead to osteoarthritis (OA) [32, 33]. Moreover, rheumatoid arthritis (RA), an autoimmune disease that results in a chronic, systemic inflammatory disorder, can lead to the destruction of articular cartilage. However, articular cartilage is avascular (does not contain blood vessels). The nutrients and oxygen are supplied to chondrocytes by diffusion which is mediated by compression of the articular cartilage or flexion of the elastic cartilage. Thus, compared to cells in other connective tissues, chondrocytes grow more slowly. The lack of vascularity and low cell metabolism restrict the self-repair ability of cartilage, ultimately leading to debilitating pain and disability [34]. OA is the most common form of arthritis, and is the leading cause of chronic disability in the United States, affecting approximately 27 million people in the United States [35].

Although current options for cartilage repair (e.g. abrasion arthroplasty, drilling, microfracture, transplantation of autograft or autologous chondrocytes) are reasonably effective to alleviate pain, these approaches have their respective limitations. For instance, drilling results in fibrocartilage of inferior quality that does not persist. Allografts suffer from lack of integration, loss of cell viability due to graft storage and concerns of disease transmission. Autografts also lack integration and require additional harvesting procedures. Consequently, repair of articular cartilage defects remains challenging [29, 36].

### 1.2.4 Cartilage Formation/Hypertrophy and Repair Process

Cartilage formation occurs early during embryonic development. First, mesenchymal cells committed to the chondrogenic fate are recruited and condense. The condensation is followed by differentiation into chondrocytes which secrete a pericellular matrix composed of characteristic ECM molecules, including collagen II and aggrecan. During embryonic development, the cartilage can serve as a template of endochondral ossification in which mature chondrocytes are flattened and form unidirectionally proliferating columns. The mature chondrocytes are progressively withdrawn from the cell cycle (prehypertrophy) and undergo hypertrophic growth. The cells undergoing hypertrophy increase in size and begin to produce a calcified matrix rich in collagen X and ALP. Hypertrophic chondrocytes also express an array of terminal differentiation genes, including metalloprotenase 13 (MMP-13), Runt-related transcription factor 2 (RunX2) and RunX3. Bone morphogenetic protein 6 (BMP-6) is found exclusively in hypertrophic chondrocytes while BMP-2 and BMP-7 can be found in pre-hypertrophic cells as well. The expression of sex-determining region Y box gene 9 (SOX-9), a transcription factor that regulates chondrogenesis in chondroprogenitor cells, is turned off in hypertrophic chondrocytes. The terminally differentiated hypertrophic chondrocytes then undergo apoptosis. This maturation process is followed by the rapid invasion of blood vessels, chondroclasts, osteoclasts and other mesenchymal cells from the perichondrium. The hypertrophic cartilage arising from the maturation process is progressively replaced by bone marrow and trabecular bone (for review, see [37]).

Articular cartilage has poor regenerative capacity, thus spontaneous repair (e.g. in the case of osteochondral fractures) is associated with defects that penetrate the underlying subchondral bone, which causes damage to the blood vessels and subsequent bleeding and hematoma formation. In such cases the resultant fibrin clot activates an inflammatory response and subsequently becomes a fibrovascular repair tissue. Various cellular components within the clot release cytokines and growth factors such as transforming growth factor β (TGF-β), platelet-derived growth factor (PDGF), insulin-like growth factors (IGFs) and BMPs to stimulate further repair. Within 2 weeks after injury, MSCs originally arising from the underlying bone marrow are recruited, proliferate and differentiate into chondrocyte-like cells and eventually a fibrocartilaginous zone is formed within the fibrin mesh. However, the matrix of the repair tissue undergoes surface fibrillation, followed by depletion of proteoglycans, chondrocyte replication and death over time. As a result, the repair tissue is replaced by a more fibrous tissue and fibrocartilage rich in collagen I. Ultimately, the repair tissue resembles a mixture of fibrocartilage and hyaline cartilage with a significant proportion of collagen I at 1 year. However, fibrocartilage does not possess the same biochemical properties as articular cartilage, and thus cannot function as normal hyaline cartilage. Consequently, the fibrocartilage eventually degenerates (for review, see [38]). The formation of such fibrocartilage is a major hurdle to articular cartilage regeneration.

## 1.3 Tissue Engineering

Although currently methods to treat bone and cartilage defects are available, there are drawbacks associated with the existing approaches. Therefore, tissue engineering has evolved to be an alternative approach to regenerating the tissues/organs.

Tissue engineering is at the interface of medicine and engineering, and generally combines the use of (1) cells, (2) scaffold and (3) biological signals, to guide the tissue regeneration. By harnessing tissue engineering approaches, new functional tissue is fabricated using cells, which can be associated with a matrix or scaffold to guide the tissue development, with the aid of signaling from biological factors [39].

### *1.3.1 Cells*

Cells are crucial for tissue regeneration as they are responsible for synthesizing ECM molecules that reshape the tissue structure and confer mechanical properties. For bone tissue engineering, osteoprogenitor cells capable of differentiating into osteoblasts (e.g. bone marrow cells [40, 41]) are commonly used, although other cell sources (e.g. fibroblasts [42]) are also exploited. In the context of cartilage engineering, chondrocytes and chondroprogenitor cells such as synoviocytes [43] and perichondrium mesenchymal cells [44] are used. Since MSCs are able to commit differentiation into the osteogenic and chondrogenic lineages, to date MSCs derived from different sources such as bone marrow (BMSCs) [45, 46] or adipose tissues (ASCs) [47, 48] are often used for the engineering of bones and cartilages. Notably, it has been suggested that BMSCs can be more easily differentiated towards the osteogenic and chondrogenic lineages than ASCs [48]. Furthermore, pluripotent stem cells such as embryonic stem cells [49, 50] and induced pluripotent stem cells [51] have been explored for tissue engineering. Due to the wide use of stem cells in tissue regeneration, regenerative medicine has been coined for stem cell technology and the terms of tissue engineering and regenerative medicine have been used interchangeably.

### *1.3.2 Scaffolds*

To develop an organized three-dimensional tissue/organ, it is important to recognize the significance of microenvironment in determining the cell's functions. For in vitro culture, the cells necessitate the scaffold for attachment and guidance of differentiation. In vivo, the cells' function is orchestrated by a symphony of signals including soluble factors, the mechanical environment (i.e. mechanical forces) and the ECM. Therefore, development of scaffold materials plays an important role in tissue engineering.

The scaffold materials can be synthetic polymers such as polyglycolic acid (PGA) [52], polylactic acid (PLA) [53], poly (L-lactide-*co*-glycolide) (PLGA) [54, 55] and poly(ethylene glycol) [56], etc. Alternatively, natural materials including agarose [57], collagen [58], GAGs [59], gelatin [60, 61], polyhydroxybutyrate (PHB) [62], hyaluronic acid (HA) [63], chitosan [64, 65] and silk [66] have been exhaustively evaluated as the scaffold materials. Since the natural ECM contains multiple matrix components, composite scaffolds consisting of multiple components are gaining popularity as well [61, 63, 67, 68]. Some of these scaffold materials, such as collagen I/III fleece (CHondro-Gide®, Cell Matrix, Sweden) and hyaluronan fleece (Hyalograft C, Fidia Advanced Biopolymers, Abano Terme, Italy), have been used in matrix-associated autologous chondrocyte transplantation (for review, see [69]). Moreover, acellular scaffolds by which the cells are removed from natural tissues have been extensively investigated [70, 71]. Since the scaffold materials and design/fabrication are beyond the scope of this Chapter, the readers may refer to previous reviews [72–74].

### *1.3.3 Biological Factors*

During bone formation/healing, a plethora of biological factors induce the migration, proliferation and differentiation of osteoprogenitor cells, and/or synthesis of matrix apposition by mature osteoblasts via the autocrine and paracrine signaling mechanisms. These factors are released to initiate bone healing and to maintain the anabolic and catabolic processes that continuously remodel bone upon bone matrix destruction. Bone injury triggers a cascade of events manifested by the influx of neutrophils, macrophages and fibroblasts. These cells express a panel of cytokines and growth factors, leading to the migration of MSCs, neovascularization, and remodeling/healing. Many growth factors, such as BMPs, IGF, TGF-β, PDGF and basic fibroblast growth factor (bFGF), can induce new bone formation through their effects on the recruitment, proliferation, and differentiation of bone-forming cells.

In particular, BMP-2 acts on global cellular mobilization and is also present during the later stages of osteogenesis and chondrogenesis, whereas BMP-7 acts on osteogenic differentiation. To date, recombinant BMP-7 (OP-1™, Stryker Corporate, Kalamazoo, Michigan) has been approved by the Food and Drug Administration (FDA) in the United States under a humanitarian device exemption for the treatment of recalcitrant long-bone non-unions and for use in revision posterolateral spinal arthrodeses [6]. Allogeneic bone graft impregnated with recombinant BMP-2 (Infuse® Bone Graft/LT-Cage®, Medtronic, Minneapolis, Minnesota) has been approved for spinal fusion [9] and recombinant human BMP-2 on an absorbable collagen sponge is approved for use in open tibia fractures [75].

In addition, it has been reported that TGF-β1 promotes osteogenic differentiation in the early and late stages of ectopic bone formation despite its inhibitory effects in vitro. Conversely, FGF-2 plays dual roles, stimulating both angiogenesis and osteogenesis, and FGF/FGF receptor signaling pathways has been suggested to coordinate bone anabolism by simultaneously activating RunX2 and BMP-2 pathways. While BMP-2 acts mainly on the osteoblastic differentiation, FGF-2 promotes cell proliferation and increases the cell population.

Bone fracture also stimulates the expression of many inflammatory cytokines including interleukin-6 (IL-6), IL-1 and TNF-α, which impart chemotactic effects on other inflammatory cells and can recruit MSCs so as to trigger the onset of the repair cascade. Whereas osteogenic growth factors are continuously expressed during bone formation and remodeling, angiogenic growth factors are predominantly expressed during the early phases to re-establish vascularity because bones are highly vascular. Numerous growth factors, including vascular endothelial growth factor (VEGF), FGF-2, PDGF, Ang-1 and IGF, have been identified to be associated with angiogenesis. During normal bone healing, VEGF expression is reported to culminate in the early phase, while BMP expression peaks at a later time point. Since establishment of a vascular bed is an early event that precedes the formation of bone, a temporal release profile mimicking the natural process may be desired to promote bone regeneration (for review, see [1]).

The cartilage repair process normally necessitates strategies that enhance the cellularity of the repair tissue, the differentiation of mesenchymal cells into chondrocytes and the production and maintenance of the cartilaginous ECM rich in collagen II and proteoglycans. As such, growth factors that support the chondrogenesis of undifferentiated mesenchymal cells or stimulate the ECM production are initial target biological factors to modulate the cellular proliferation and differentiation. The growth factors interact specifically with their membrane-bound receptors and trigger downstream signaling pathway, leading to the induction of response genes controlling cell proliferation and differentiation. For instance, TGF-β2 mediates hypertrophic differentiation of chondrocytes by regulating Indian hedgehog (Ihh) and parathormone-related peptide (PTHrP) expression [76]. Members of the TGF-β superfamily such as TGF-β1, TGF-β2, BMP-2, IGF-1 are released and sequestered in the ECM and have been suggested to be associated with cell proliferation and cartilage homeostasis [77]. TGF-β1 directly stimulates proteoglycan and collagen synthesis [78] and antagonizes the effects of IL-1 on matrix metalloproteinase in normal and osteoarthritic chondrocytes [79]. IGF-1 induces cartilage proteoglycan synthesis and collagen matrix production in vitro and in vivo [80]. BMP-2 stimulates chondrocyte production of proteoglycan [81] and induces the expression of chondrogenesis associated proteins (e.g. SOX-9, collagen II and aggrecan) in synovium-derived progenitor cells cultured in three-dimensional alginate hydrogel [82]. Other factors such as BMP-7 [83] and cartilage-derived morphogenetic protein-1 (CDMP-1/GDF-5) [84] are found in the embryonic limb bud, the fetal growth plate and the joint interzones of developing limbs. These factors have been shown to induce chondrogenic differentiation in vitro [85, 86]. For instance, BMP-7 increases cartilage-specific matrix synthesis in articular chondrocytes in vitro and enhances the healing of osteochondral defects in vivo [87]. Furthermore, BMP-7 promotes cartilage differentiation and protects engineered cartilage from fibroblast invasion and destruction [83]. CDMP-2 [81] and FGF-2 also stimulate chondrocyte production of proteoglycan. Conversely, PTHrP hinders terminal differentiation of cultured chondrocytes and stimulates aggrecan and collagen II synthesis [88, 89].

In addition to the aforementioned secreted factors, transcription factors such as SOX9 [90] and Cbfa-1/Runx-2 [91] as well as signaling molecules such as Wnt [92] and hedgehog [93] have been implicated in chondrogenesis. Furthermore, the regulation of cartilage homeostasis and function relies on the intricate balance between anabolic and catabolic processes, therefore besides the factors that enhance anabolism, inhibitors of catabolic programs have also been assessed. Potential targets include such cytokines as tumor necrosis factor (TNF), IL-1 and IL-17 (for review, see [94]).

## References


1. Santo VE, Gomes ME, Mano JF, Reis RL (2013) Controlled release strategies for bone, cartilage, and osteochondral engineering-Part I: recapitulation of native tissue healing and variables for the design of delivery systems. Tissue Eng Part B Rev 19:308–326
2. Buckwalter JA, Einhorn TA, Bolander ME, Cruess RL (1996) Healing of the musculoskeletal tissues. In: Bucholz RW, Heckman JD, Court-Brown C, Tornetta P, Koval KJ, Wirth MA (eds) 4th edn. Lippincott Williams & Wilkins, Philadelphia, pp 261–304
3. Ross MH, Reith EJ, Romrell LJ (1989) Histology: a text and atlas. Williams & Wilkins, Baltimore
4. Bueno EM, Glowacki J (2009) Cell-free and cell-based approaches for bone regeneration. Nat Rev Rheumatol 5:685–697
5. Gilbert SF (2000) Developmental biology. Sinauer Associates, Sunderland
6. Carofino BC, Lieberman JR (2008) Gene therapy applications for fracture-healing. J Bone Joint Surg Am 90A:99–110
7. Phillips JE, Gersbach CA, Garcia AJ (2007) Virus-based gene therapy strategies for bone regeneration. Biomaterials 28:211–229
8. Pelled G, Ben-Arav A, Hock C, Reynolds DG, Yazici C, Zilberman Y et al (2010) Direct gene therapy for bone regeneration: gene delivery, animal models, and outcome measures. Tissue Eng Part B Rev 16:13–20
9. Kimelman N, Pelled G, Helm GA, Huard J, Schwarz EM, Gazit D (2007) Review: gene- and stem cell-based therapeutics for bone regeneration and repair. Tissue Eng 13:1135–1150
10. Zimmermann G, Wagner C, Schmeckenbecher K, Wentzensen A, Moghaddam A (2009) Treatment of tibial shaft non-unions: bone morphogenetic proteins versus autologous bone graft. Injury 40(Suppl 3):S50–S53
11. Tseng SS, Lee MA, Reddi H (2008) Nonunions and the potential of stem cells in fracture-healing. J Bone Joint Surg Am 90A:92–98
12. Marino JT, Ziran BH (2010) Use of solid and cancellous autologous bone graft for fractures and nonunions. Orthop Clin North Am 41:15–26
13. Kneser U, Schaefer DJ, Polykandriotis E, Horch RE (2006) Tissue engineering of bone: the reconstructive surgeon's point of view. J Cell Mol Med 10:7–19
14. Gamradt SC, Lieberman JR (2004) Genetic modification of stem cells to enhance bone repair. Ann Biomed Eng 32:136–147
15. Eward WC, Kontogeorgakos V, Levin LS, Brigman BE (2010) Free vascularized fibular graft reconstruction of large skeletal defects after tumor resection. Clin Orthop Relat Res 468:590–598
16. Smith JO, Aarvold A, Tayton ER, Dunlop DG, Oreffo RO (2011) Skeletal tissue regeneration: current approaches, challenges, and novel reconstructive strategies for an aging population. Tissue Eng Part B Rev 17:307–320
17. Hernigou P, Poignard A, Manicom O, Mathieu G, Rouard H (2005) The use of percutaneous autologous bone marrow transplantation in nonunion and avascular necrosis of bone. J Bone Joint Surg Br 87:896–902
18. Szpalski C, Barr J, Wetterau M, Saadeh PB, Warren SM (2010) Cranial bone defects: current and future strategies. Neurosurg Focus 29:E8
19. Bose S, Roy M, Bandyopadhyay A (2012) Recent advances in bone tissue engineering scaffolds. Trends Biotechnol 30:546–554

20. Shegarfi H, Reikeras O (2009) Bone transplantation and immune response. J Orthop Surg 17:206–211
21. Owen TA, Aronow M, Shalhoub V, Barone LM, Wilming L, Tassinari MS et al (1990) Progressive development of the rat osteoblast phenotype in vitro – reciprocal relationships in expression of genes associated with osteoblast proliferation and differentiation during formation of the bone extracellular matrix. J Cell Physiol 143:420–430
22. Liu Y, Wang L, Kikuiri T, Akiyama K, Chen C, Xu X et al (2012) Mesenchymal stem cell-based tissue regeneration is governed by recipient T lymphocytes via IFN-γ and TNF-α. Nat Med 17:1594–1601
23. Santos MI, Reis RL (2010) Vascularization in bone tissue engineering: physiology, current strategies, major hurdles and future challenges. Macromol Biosci 10:12–27
24. Hankenson KD, Dishowitz M, Gray C, Schenker M (2011) Angiogenesis in bone regeneration. Injury 42:556–561
25. Trippel SB, Ghivizzani SC, Nixon AJ (2004) Gene-based approaches for the repair of articular cartilage. Gene Ther 11:351–359
26. Knudson CB, Knudson W (2001) Cartilage proteoglycans. Semin Cell Dev Biol 12:69–78
27. Grayson WL, Chao P-HG, Marolt D, Kaplan DL, Vunjak-Novakovic G (2008) Engineering custom-designed osteochondral tissue grafts. Trends Biotechnol 26:181–189
28. Barbero A, Grogan S, Schafer D, Heberer M, Mainil-Varlet P, Martin I (2004) Age related changes in human articular chondrocyte yield, proliferation and post-expansion chondrogenic capacity. Osteoarthritis Cartilage 12:476–484
29. Huey DJ, Hu JC, Athanasiou KA (2012) Unlike bone, cartilage regeneration remains elusive. Science 338:917–921
30. Gannon AR, Nagel T, Kelly DJ (2012) The role of the superficial region in determining the dynamic properties of articular cartilage. Osteoarthritis Cartilage 20:1417–1425
31. Sophia Fox AJ, Bedi A, Rodeo SA (2009) The basic science of articular cartilage: structure, composition, and function. Sports Health 1:461–468
32. Sandell LJ, Aigner T (2001) Articular cartilage and changes in arthritis. An introduction: cell biology of osteoarthritis. Arthritis Res 3:107–113
33. Bock HC, Michaeli P, Bode C, Schultz W, Kresse H, Herken R et al (2001) The small proteoglycans decorin and biglycan in human articular cartilage of late-stage osteoarthritis. Osteoarthritis Cartilage 9:654–663
34. Vinatier C, Mrugala D, Jorgensen C, Guicheux J, Noel D (2009) Cartilage engineering: a crucial combination of cells, biomaterials and biofactors. Trends Biotechnol 27:307–314
35. Van Manen MD, Nace J, Mont MA (2012) Management of primary knee osteoarthritis and indications for total knee arthroplasty for general practitioners. J Am Osteopath Assoc 112:709–715
36. Cui L, Wu Y, Cen L, Zhou H, Yin S, Liu G et al (2009) Repair of articular cartilage defect in non-weight bearing areas using adipose derived stem cells loaded polyglycolic acid mesh. Biomaterials 30:2683–2693
37. Kirn-Safran CB, Gomes RR, Brown AJ, Carson DD (2004) Heparan sulfate proteoglycans: coordinators of multiple signaling pathways during chondrogenesis. Birth Defects Res C Embryo Today 72:69–88
38. Tang QO, Shakib K, Heliotis M, Tsiridis E, Mantalaris A, Ripamonti U (2009) TGF-beta3: a potential biological therapy for enhancing chondrogenesis. Expert Opin Biol Ther 9:689–701
39. Langer R, Vacanti JP (1993) Tissue engineering. Science 260:920–926
40. Feeley BT, Conduah AH, Sugiyama O, Krenek L, Chen ISY, Lieberman JR (2006) In vivo molecular imaging of adenoviral versus lentiviral gene therapy in two bone formation models. J Orthop Res 24:1709–1721
41. Virk MS, Conduah A, Park SH, Liu N, Sugiyama O, Cuomo A et al (2008) Influence of short-term adenoviral vector and prolonged lentiviral vector mediated bone morphogenetic protein-2 expression on the quality of bone repair in a rat femoral defect model. Bone 42:921–931
42. Lattanzi W, Parrilla C, Fetoni A, Logroscino G, Straface G, Pecorini G et al (2008) Ex vivo-transduced autologous skin fibroblasts expressing human Lim mineralization protein-3 efficiently form new bone in animal models. Gene Ther 15:1330–1343

43. Zhang XL, Mao ZB, Yu CL (2004) Suppression of early experimental osteoarthritis by gene transfer of interleukin-1 receptor antagonist and interleukin-10. J Orthop Res 22:742–750
44. Gelse K, von der Mark K, Aigner T, Park J, Schneider H (2003) Articular cartilage repair by gene therapy using growth factor- producing mesenchymal cells. Arthritis Rheum 48:430–441
45. Cao L, Yang F, Liu G, Yu D, Li H, Fan Q et al (2011) The promotion of cartilage defect repair using adenovirus mediated Sox9 gene transfer of rabbit bone marrow mesenchymal stem cells. Biomaterials 32:3910–3920
46. Sheyn D, Ruthemann M, Mizrahi O, Kallai I, Zilberman Y, Tawackoli W et al (2010) Genetically modified mesenchymal stem cells induce mechanically stable posterior spine fusion. Tissue Eng Part A 16:3679–3686
47. Lee J-M, Im G-I (2012) SOX trio-co-transduced adipose stem cells in fibrin gel to enhance cartilage repair and delay the progression of osteoarthritis in the rat. Biomaterials 33:2016–2024
48. Lin C-Y, Lin K-J, Kao C-Y, Chen M-C, Yen T-Z, Lo W-H et al (2011) The role of adipose-derived stem cells engineered with the persistently expressing hybrid baculovirus in the healing of massive bone defects. Biomaterials 32:6505–6514
49. Jukes JM, Moroni L, van Blitterswijk CA, de Boer J (2008) Critical steps toward a tissue-engineered cartilage implant using embryonic stem cells. Tissue Eng Part A 14:135–147
50. Toh WS, Lee EH, Guo X-M, Chan JKY, Yeow CH, Choo AB et al (2010) Cartilage repair using hyaluronan hydrogel-encapsulated human embryonic stem cell-derived chondrogenic cells. Biomaterials 31:6968–6980
51. Ye J-H, Xu Y-J, Gao J, Yan S-G, Zhao J, Tu Q et al (2011) Critical-size calvarial bone defects healing in a mouse model with silk scaffolds and SATB2-modified iPSCs. Biomaterials 32:5065–5076
52. Mahmoudifar N, Doran PM (2010) Chondrogenic differentiation of human adipose-derived stem cells in polyglycolic acid mesh scaffolds under dynamic culture conditions. Biomaterials 31:3858–3867
53. Zhou C, Shi Q, Guo W, Terrell L, Qureshi AT, Hayes DJ et al (2013) Electrospun bio-nanocomposite scaffolds for bone tissue engineering by cellulose nanocrystals reinforcing maleic anhydride grafted PLA. ACS Appl Mater Interf 5:3847–3854
54. Chen H-C, Chang Y-H, Chuang C-K, Lin C-Y, Sung L-Y, Wang Y-H et al (2009) The repair of osteochondral defects using baculovirus-mediated gene transfer with de-differentiated chondrocytes in bioreactor culture. Biomaterials 30:674–681
55. Wang W, Li B, Li Y, Jiang Y, Ouyang H, Gao C (2010) In vivo restoration of full-thickness cartilage defects by poly(lactide-co-glycolide) sponges filled with fibrin gel, bone marrow mesenchymal stem cells and DNA complexes. Biomaterials 31:5953–5965
56. Williams CG, Kim TK, Taboas A, Malik A, Manson P, Elisseeff J (2003) In vitro chondrogenesis of bone marrow-derived mesenchymal stem cells in a photopolymerizing hydrogel. Tissue Eng 9:679–688
57. Thorpe SD, Buckley CT, Vinardell T, O'Brien FJ, Campbell VA, Kelly DJ (2010) The response of bone marrow-derived mesenchymal stem cells to dynamic compression following TGF-beta3 induced chondrogenic differentiation. Ann Biomed Eng 38:2896–2909
58. Bright C, Park YS, Sieber AN, Kostuik JP, Leong KW (2006) In vivo evaluation of plasmid DNA encoding OP-1 protein for spine fusion. Spine (Phila Pa 1976) 31:2163–2172
59. Deng T, Lv J, Pang J, Liu B, Ke J (2012) Construction of tissue-engineered osteochondral composites and repair of large joint defects in rabbit. J Tissue Eng Regen Med. doi:10.1002/term.556
60. Awad HA, Wickham MQ, Leddy HA, Gimble JM, Guilak F (2004) Chondrogenic differentiation of adipose-derived adult stem cells in agarose, alginate, and gelatin scaffolds. Biomaterials 25:3211–3222
61. Chang CH, Kuo TF, Lin CC, Chou CH, Chen KH, Lin FH et al (2006) Tissue engineering-based cartilage repair with allogenous chondrocytes and gelatin-chondroitin-hyaluronan tri-copolymer scaffold: a porcine model assessed at 18, 24, and 36 weeks. Biomaterials 27:1876–1888
62. Ye C, Hu P, Ma MX, Xiang Y, Liu RG, Shang XW (2009) PHB/PHBHHx scaffolds and human adipose-derived stem cells for cartilage tissue engineering. Biomaterials 30:4401–4406

63. Fan H, Tao H, Wu Y, Hu Y, Yan Y, Luo Z (2010) TGF-β3 immobilized PLGA-gelatin/chondroitin sulfate/hyaluronic acid hybrid scaffold for cartilage regeneration. J Biomed Mater Res A 95:982–992
64. Nettles DL, Elder SH, Gilbert JA (2002) Potential use of chitosan as a cell scaffold material for cartilage tissue engineering. Tissue Eng 8:1009–1016
65. Wang XH, Cui FZ, Feng QL, Li JC, Zhang YH (2003) Preparation and characterization of collagen/chitosan matrices as potential biomaterials. J Bioact Compat Polym 18:453–467
66. Marolt D, Augst A, Freed LE, Vepari C, Fajardo R, Patel N et al (2006) Bone and cartilage tissue constructs grown using human bone marrow stromal cells, silk scaffolds and rotating bioreactors. Biomaterials 27:6138–6149
67. Li ZS, Zhang MQ (2005) Chitosan-alginate as scaffolding material for cartilage tissue engineering. J Biomed Mater Res 75A:485–493
68. Zhao L, Chang J (2004) Preparation and characterization of macroporous chitosan/wollastonite composite scaffolds for tissue engineering. J Mater Sci Mater Med 15:625–629
69. Hildner F, Albrecht C, Gabriel C, Redl H, van Griensven M (2011) State of the art and future perspectives of articular cartilage regeneration: a focus on adipose-derived stem cells and platelet-derived products. J Tissue Eng Regen Med 5:e36–e51
70. Cheng NC, Estes BT, Awad HA, Guilak F (2009) Chondrogenic differentiation of adipose-derived adult stem cells by a porous scaffold derived from native articular cartilage extracellular matrix. Tissue Eng Part A 15:231–241
71. Steinhoff G, Stock U, Karim N, Mertsching H, Timke A, Meliss RR et al (2000) Tissue engineering of pulmonary heart valves on allogenic acellular matrix conduits: in vivo restoration of valve tissue. Circulation 102:III-50–III-55
72. Carletti E, Motta A, Migliaresi C (2011) Scaffolds for tissue engineering and 3D cell culture. In: Haycock JW (ed) 3D cell culture, vol 695. Humana Press, New York, pp 17–39
73. Filardo G, Kon E, Roffi A, Di Martino A, Marcacci M (2013) Scaffold-based repair for cartilage healing: a systematic review and technical note. Arthroscopy 29:174–186
74. Zanetti AS, Sabliov C, Gimble JM, Hayes DJ (2013) Human adipose-derived stem cells and three-dimensional scaffold constructs: a review of the biomaterials and models currently used for bone regeneration. J Biomed Mater Res B Appl Biomater 101:187–199
75. Boden SD (2005) The ABCs of BMPs. Orthop Nurs 24:49–52
76. Alvarez J, Sohn P, Zeng X, Doetschman T, Robbins DJ, Serra R (2002) TGFbeta2 mediates the effects of hedgehog on hypertrophic differentiation and PTHrP expression. Development 129:1913–1924
77. van der Kraan PM, Buma P, van Kuppevelt T, van den Berg WB (2002) Interaction of chondrocytes, extracellular matrix and growth factors: relevance for articular cartilage tissue engineering. Osteoarthritis Cartilage 10:631–637
78. Redini F, Galera P, Mauviel A, Loyau G, Pujol JP (1988) Transforming growth factor-β1 stimulates collagen and glycosaminoglycan biosynthesis in cultured rabbit articular chondrocytes. FEBS Lett 234:172–176
79. Andrews HJ, Edwards TA, Cawston TE, Hazleman BL (1989) Transforming growth factor-β1 causes partial inhibition of interleukin 1-stimulated cartilage degradation in vitro. Biochem Biophys Res Commun 162:144–150
80. Madry H, Padera R, Seidel J, Langer R, Freed LE, Trippel SB et al (2002) Gene transfer of a human insulin-like growth factor I cDNA enhances tissue engineering of cartilage. Hum Gene Ther 13:1621–1630
81. Li J, Kim KS, Park JS, Elmer WA, Hutton WC, Yoon ST (2003) BMP-2 and CDMP-2: stimulation of chondrocyte production of proteoglycan. J Orthop Sci 8:829–835
82. Park Y, Sugimoto M, Watrin A, Chiquet M, Hunziker EB (2005) BMP-2 induces the expression of chondrocyte-specific genes in bovine synovium-derived progenitor cells cultured in three-dimensional alginate hydrogel. Osteoarthritis Cartilage 13:527–536
83. Kaps C, Bramlage C, Smolian H, Haisch A, Ungethum U, Burmester GR et al (2002) Bone morphogenetic proteins promote cartilage differentiation and protect engineered artificial cartilage from fibroblast invasion and destruction. Arthritis Rheum 46:149–162

84. Merino R, Macias D, Ganan Y, Economides AN, Wang X, Wu Q et al (1999) Expression and function of GDF-5 during digit skeletogenesis in the embryonic chick leg bud. Dev Biol 206:33–45
85. Gruber R, Mayer C, Bobacz K, Krauth MT, Graninger W, Luyten FP et al (2001) Effects of cartilage-derived morphogenetic proteins and osteogenic protein-1 on osteochondrogenic differentiation of periosteum-derived cells. Endocrinology 142:2087–2094
86. Klein-Nulend J, Semeins CM, Mulder JW, Winters HAH, Goei SW, Ooms ME et al (1998) Stimulation of cartilage differentiation by osteogenic protein-1 in cultures of human perichondrium. Tissue Eng 4:305–313
87. Louwerse RT, Heyligers IC, Klein-Nulend J, Sugihara S, van Kampen GP, Semeins CM et al (2000) Use of recombinant osteogenic protein-1 for the repair of subchondral defects in articular cartilage in goats. J Biomed Mater Res 49:506–516
88. Harvey AK, Yu XP, Frolik CA, Chandrasekhar S (1999) Parathyroid hormone-(1–34) enhances aggrecan synthesis via an insulin-like growth factor-I pathway. J Biol Chem 274:23249–23255
89. Erdmann S, Muller W, Bahrami S, Vornehm SI, Mayer H, Bruckner P et al (1996) Differential effects of parathyroid hormone fragments on collagen gene expression in chondrocytes. J Cell Biol 135:1179–1191
90. Bi WM, Deng JM, Zhang ZP, Behringer RR, de Crombrugghe B (1999) Sox9 is required for cartilage formation. Nat Genet 22:85–89
91. Inada M, Yasui T, Nomura S, Miyake S, Deguchi K, Himeno M et al (1999) Maturational disturbance of chondrocytes in Cbfa1-deficient mice. Dev Dyn 214:279–290
92. Hartmann C, Tabin CJ (2000) Dual roles of Wnt signaling during chondrogenesis in the chicken limb. Development 127:3141–3159
93. Vortkamp A, Lee K, Lanske B, Segre GV, Kronenberg HM, Tabin CJ (1996) Regulation of rate of cartilage differentiation by Indian hedgehog and PTH-related protein. Science 273:613–622
94. Evans CH, Gouze E, Gouze JN, Robbins PD, Ghivizzani SC (2006) Gene therapeutic approaches–transfer in vivo. Adv Drug Deliv Rev 58:243–258

# Chapter 2
# Viral Gene Therapy Vectors

**Abstract** The vectors for gene delivery can be divided into two classes: viral and nonviral. Despite the rapid progress in the development of nonviral gene delivery, viral vectors such as retrovirus/lentivirus, adenovirus, adeno-associated virus (AAV) and baculovirus mediate more efficient delivery than nonviral vectors, especially for primary cells. This chapter briefly compares the nonviral and viral vectors and mainly discusses the development and characteristics of these viral vectors.

## 2.1 Gene Therapy

Gene therapy is a technique originally developed to deliver DNA or RNA molecules to cells/tissues for the treatment of genetic diseases. Nonetheless, gene therapy has been rapidly expanded to a wide variety of applications such as treatment of cancers and infectious diseases. For example, delivery of Diphtheria toxin A gene via baculovirus inhibits the glioma xenograft growth in the rat brain [1] and delivery of the hemagglutinin gene of avian influenza virus to mice and poultry via adenovirus elicits immunity and confers protection [2]. Gene therapy and tissue engineering have also converged for the repair of various tissues/organs, such as musculoskeletal and cardiovascular systems. For instance, gene therapy in conjunction with tissue engineering can aid in the treatment of myocardial infarction [3, 4], bone defects [5] and diseases in cartilage [6–8]. Although protein-based therapy also provides an effective approach for the treatment of bone/cartilage defects, it might be ineffective in repairing large defects clinically [9]. In the case of bone repair, a single protein (e.g. bone morphogenetic protein 2 (BMP-2)) dose may not confer an adequate

Y.-C. Hu, *Gene Therapy for Cartilage and Bone Tissue Engineering*,
SpringerBriefs in Bioengineering, DOI 10.1007/978-3-642-53923-7_2,

repair response because of the short protein half-life and poor retention in large defects [10], thus milligrams of proteins or multiple doses are required. With this regard, gene therapy offers an attractive option to augment tissue repair [10]. Since the gene, rather than a degradable protein, is being delivered, gene delivery potentially results in higher and more sustained protein release in a more physiologic manner than recombinant protein therapy [10]. Moreover, the endogenously synthesized proteins may have greater biological effectiveness than their exogenous counterpart.

Gene therapy can be performed either in vivo or ex vivo. The in vivo gene delivery involves the injection or implantation of genetic material carried by the delivery vector directly into the host. This approach is simpler and minimizes the risk of infection since only one procedure is required [11]. However, direct vector injection, either locally or systemically, may elicit inflammatory responses which interfere with the reparative process [12]. It is also extremely difficult to specifically deliver the genes into target cells in vivo, thus resulting in low levels of protein expression. In the context of tissue engineering, typically the vector is administered locally to minimize unwanted side effects, but it is difficult to avoid the transgene expression in secondary tissues. Another challenge is how to achieve a sustained long-term expression of the therapeutic gene, although in some cases a short-term expression is sufficient to accelerate healing of tissues and may be desirable [10].

Ex vivo gene delivery involves the genetic modification of cells (either autogenic or allogenic) and re-introduction into the host with or without a scaffold. The ex vivo strategy enhances the roles of the cells in the regenerative process with autocrine/paracrine effects from the expressed transgene products. In one ex vivo gene therapy study, the articular chondrocytes are genetically modified by transfection of cDNA encoding insulin-like growth factor 1 (IGF-1), encapsulated and implanted into osteochondral defects in rabbits. The genetic modification results in prolonged IGF-1 expression in vitro (up to 36 days), augments articular cartilage repair and accelerates the formation of subchondral bone [13]. The disadvantage of such ex vivo therapy in the tissue engineering setting is that it involves two separate invasive procedures for a patient (when autologous cells are used), which increases the pain the patients suffer and possibility of morbidity. Moreover, the cells transplanted into the defects may not persist for a period of time sufficient to heal the defects [14].

## 2.2 Vectors for Gene Delivery

### *2.2.1 Nonviral vs. Viral Vectors*

The vectors for gene delivery can be divided into two classes: viral and nonviral. Nonviral vectors mainly rely on the delivery of plasmid DNA (or other forms of DNA/RNA) into cells/tissues with the aid of a proper transfection reagent (e.g. liposomes, polymer-based molecular conjugates, nanoparticles, etc). The nonviral vector carrying sex-determining region Y box gene 9 (SOX-9) gene has been shown to enhance the chondrogenesis of mouse bone marrow-derived mesenchymal stem cells (BMSCs) [15]. Transfection of the DNA encoding IGF-1 into articular chondrocytes and transplantation of transfected cells also lead to the formation of a new tissue layer on the cartilage explant surface [16]. In addition, primary chondrocytes and explants can be engineered by transfection of DNA encoding human glycosaminoglycan (GAG)-synthesizing enzyme, β1,3-glucuronosyltransferase-I (GlcAT-1). Such GlcAT-1 delivery enhances the glycosaminoglycan (GAG) deposition and overcomes interleukin 1β (IL-1β)-induced proteoglycan depletion [17].

Despite the rapid progress in the development of nonviral gene delivery vector, transfection methods (e.g. in vivo electroporation [18], microporation [19], nucleofection [20]) and transfection reagents (e.g. FuGENE6 [21], nanoparticles [22, 23]) to enhance the efficiency of transfection into cells, it is generally perceived that the efficiency of gene delivery mediated by nonviral vectors is lower when compared with that by viral vectors [24–26]. In particular, transfection of adult MSCs is very inefficient [27].

In contrast, viral vectors are widely employed for gene therapy as viruses naturally evolve mechanisms for effective delivery of their genetic cargo into cells for replication and expression [28]. For vector development, generally the elements of viral genome that contribute to replication, virulence and pathology are deleted and replaced by gene(s) of interest while retaining the elements contributing to efficient delivery [24, 29]. The viral vectors that are in common use include retrovirus/lentivirus, adenovirus and adeno-associated virus (AAV) [25, 26]. Some emerging viral vectors such as baculovirus have also been investigated (for review, see [30–35]). Therefore, this chapter mainly deals with these viral vectors.

### 2.2.2 *Retrovirus/Lentivirus*

Retrovirus genome comprises two identical RNA molecules, which after entry into the cells are reverse transcribed to complementary DNA and integrate into the host chromosome. The integration ensures the persistence of the therapeutic gene in the cells, thus retrovirus is initially favored for applications whereby long-term expression is desired. In fact, retrovirus is the viral vector used for the first human gene therapy clinical trial for the treatment of a genetic disease known as severe combined immunodeficiency (SCID) [36]. In this clinical trial, a gene encoding adenosine deaminase (ADA) is transduced into autologous T cells using the retroviral vector and delivered into two patients. Four years after the initial treatment, the ADA expression is still detectable and the symptoms are alleviated in one patient, thus demonstrating the proof-of concept of gene therapy [36].

The first and the most commonly used retrovirus is Molony murine leukemia virus (MuLV). MuLV transduces synovial fibroblasts cultured in vitro with reasonable efficiency, but is inactive in in vivo experiments when injected into the knee joints. The inefficiency is partly due to the fact that retrovirus only transduces proliferating cells, but not quiescent cells. Besides, the extracellular matrix (ECM) hinders retrovirus from directly transducing chondrocytes embedded in the ECM. Therefore, retrovirus is more suitable for ex vivo gene transfer-mediated tissue engineering. Retrovirus expressing enhanced green fluorescent protein (EGFP) is used to transduce chondrocytes, followed by implantation of the transduced cells into full-thickness defects in knee joints of rabbits [37]. The EGFP expression and the number of implanted chondrocytes remain stable for at least 4 weeks in vivo.

Note, however, that the integration does not guarantee long-term transgene expression because (1) the genes might be silenced as a result of epigenetic modification [38] and (2) the transduced cells may be eradicated due to regular turnover or the immune system. Retrovirus-mediated transfer to autologous synoviocytes results in transgene expression that steadily diminishes over a period of 4–6 weeks following intra-articular implantation of the transduced cells [39].

Moreover, retrovirus transduces synoviocytes in the inflamed joints better than those in the naïve joints [8]. Since one of the primary symptoms of rheumatoid arthritis (RA) is the thickening of the synovium through synovial cell proliferation, the study implicates the application of retroviral vectors in gene delivery to the joints of RA patients [8]. Retroviral vector-based gene therapy for RA treatment has entered clinical trial [6], in which the retrovirus expressing interleukin 1 (IL-1) receptor antagonist (IL-1Ra) is used to transduce autologous synovial cells ex vivo, followed by implantation of the transduced cells into metacarpophalangeal joints of RA patients 1 week prior to the scheduled joint replacement surgery. No adverse effects related to the gene transfer are observed and there is no relevant spread of the transgene to extra-articular sites [6].

However, retrovirus preferentially integrates viral genes to the transcription start sites and highly expressed genes in the host chromosome, thus raising serious safety concerns. Maria Cavazzana-Calvo, Alain Fischer and coworkers have demonstrated the use of retroviral vectors for the cure of X-linked SCID in nine out ten patients [40]. Unfortunately, two of the patients develop leukemia as retrovirus integrates into chromosomal sites in proximity to the LMO-2 proto-oncogene promoter, leading to aberrant expression of LMO2 [41]. Follow-up studies have cured more than 20 patients and confirmed the efficacy of SCID gene therapy, but leukemia occurs in several more patients [42–44]. Furthermore, the in vivo application of retrovirus in humans is compromised by their sensitivity to inactivation by the complement system [45].

Lentivirus belongs to the retrovirus family but is different in that lentivirus has a more complex genome and is capable of transducing non-dividing cells. Similar to other retroviruses, lentivirus has a low DNA carrying capacity of $\approx$8 kb and can stably integrate the viral genes into the host genome. The transduction spectrum of lentivirus can be broadened by pseudotyping the virus envelope with VSV-G (vesicular stomatitis virus G protein). Such VSVG-pseudotyped lentivirus can transduce cultured chondrocytes and mediate gene transfer to synovium [46]. Lentivirus expressing SOX-9 also enhances collagen II expression and down-regulates the collagen I expression of passaged chondrocytes, implicating the potential of lentivirus-mediated SOX-9 expression in restoring chondrocyte phenotype even after de-differentiation [47]. Lentivirus is also used for intra-articular delivery of endostatin [48] and angiostatin [49] into rodents to treat experimental models of RA. To date, use of lentivirus for ex vivo transduction of $CD34^+$ cells has entered into clinical trials. One example is the use of lentiviral vector for the treatment of patients with X-linked adrenoleukodystrophy (ALD). The trial confirms that lentivirus-mediated gene therapy provides clinical benefits in ALD [50].

However, lentiviral vectors (e.g. the vector derived from human immunodeficiency virus (HIV)) also favor the integration into active transcription units [51]. The integration may elicit insertional mutagenesis [52, 53] and raise safety concerns. To circumvent these problems, non-integrating lentiviral vectors have been developed [54]. These lentiviral vectors are defective in integrase (the enzyme responsible for integration), thus enabling the maintenance of transgene in the episomal form while conferring stable transgene expression [55].

Unlike the popular use of retrovirus/lentivirus in other applications requiring long-term expression (e.g. treatment of inherited diseases), whether retroviral/lentiviral vectors will be deemed safe for clinical use in bone/cartilage tissue engineering is questionable. Justification of the safe use of retroviral/lentiviral vectors in bone/cartilage regeneration requires further testing.

### 2.2.3 *Adenovirus*

Adenovirus has a 36 kb, double stranded DNA genome packaged in a 100 nm icosahedral capsid. Wild-type adenovirus infects cells in the upper respiratory tract and can result in mild cold. Adenovirus is able to infect dividing and nondividing cells, which provides advantages for in vivo gene delivery (for review, see [56]). There are more than 50 adenovirus serotypes and initially emphasis is placed on serotype 5 (Ad5). Nonetheless, more and more adenovirus serotypes are explored for gene therapy.

The first generation adenoviral vectors are deleted in E1 and E3 genes, which allows for the insertion of up to 8 kb of foreign gene cassette. Adenovirus can be produced to high titers, which renders the vector production simpler and more cost-effective. Adenovirus genome does not integrate at high efficiency and remains episomal, thus the viral genomes only persist in non-dividing cells and the therapeutic gene expression is lost when the transduced cells are gradually diluted out of the population.

The adenovirus expressing the *lacZ* gene is employed for in vivo gene delivery to joints, which results in transgene expression for over a month within the synovium without provoking an inflammatory response [57]. However, inflammatory responses are found in subsequent studies exploiting adenovirus-mediated delivery of p53 [58], IL-1 and tumor necrosis factor $\alpha$ (TNF-$\alpha$) receptors [59] to disease joints, causing the rapid decline and extinguishing of transgene expression in 2–4 weeks. The transient expression results from the strong cellular immune responses elicited by the continued expression of endogenous viral proteins within the cells.

Nonetheless, adenoviral vector is widely used for gene therapy in the context of tissue engineering. In particular, the adenovirus-mediated IGF-1 expression can last for at least 28 days and effectively enhances in vitro chondrogenesis [60]. Adenovirus-mediated IGF-1, transforming growth factor β1 (TGF-β1) and BMP-2 expression in chondrocytes greatly increases matrix synthesis in vitro, even in the presence of the inflammatory cytokine IL-1 [61]. Furthermore, two adenoviral vectors expressing IGF-1 and IL-1Ra are used in combination to co-transduce cultured synoviocytes. The IGF-1 and IL-1Ra secreted by the transduced cells fully reverse the depletion of cartilage proteoglycan contents induced by IL-1 [62].

The biggest barrier to the clinical application of adenovirus is the strong humoral and cellular immune responses it elicits, especially after the tragedy death in an adenovirus-mediated gene therapy trial [63]. To minimize the immune responses, the "gutless" vectors that contain only the viral terminal repeats and the packaging sequence are developed [64]. In the gutless vector, all other viral components are deleted, thus it can accommodate up to 36 kb of exogenous DNA and does not trigger strong immune responses. However, gutless adenovirus vector, due to the deletion of most viral components, requires helper plasmid or virus for production, making the production process more complicated. Furthermore, the transgene expression appears to be weakened [65]. Additionally, most humans have pre-existing immunity to adenovirus which could neutralize the administered virus vectors. Such pre-existing immunity problem may be circumvented by using adenoviral vectors derived from different serotypes.

### *2.2.4 AAV*

AAV is a parvovirus with a ≈4,700 nt, single-stranded DNA genome (for review, see [66]). The genome replication of AAV requires helper viruses such as adenovirus or herpes simplex virus to provide helper functions. AAV alone is non-pathogenic to humans and does not induce serious host immune responses. AAV can mediate long-term transgene expression in a wide variety of cells, including dividing and non-dividing cells. These advantages have inspired the wide application of recombinant AAV vectors for gene delivery [67]. Like adenovirus, AAV exists in many serotypes, among them AAV 2 and AAV 5 are most extensively studied and utilized.

The feasibility of utilizing AAV vector in tissue engineering has been demonstrated by a study in which AAV is used for direct in vivo modification of synoviocytes and the β-galactosidase expression in the synovium is observed for at least 7 months [68]. A subsequent study also shows efficient (transduction efficiency >70 %) and persistent recombinant AAV vector transduction of chondrocytes derived from normal and osteoarthritic human articular cartilage [69]. Strikingly, transduction of explant cultures of articular cartilage results in reporter gene expression within the tissue to a depth exceeding 450 μm, which remains persistent for 150 days [69]. These data suggest that AAV vectors are able to transduce chondrocytes in situ within their native matrix to a depth sufficient to be of important clinical significance [69].

Furthermore, AAV-mediated delivery of TGF-β1 gene improves the expression of collagen II and aggrecan while decreases the matrix metalloproteinase 3 (MMP-3) expression in cultured normal and osteoarthritic chondrocytes [70]. In vitro transduction of chondrocytes with an AAV vector expressing fibroblast growth factor (FGF-2) stimulates cell proliferation over a long period of time, and in vivo application of the same AAV vector significantly improves the overall repair of osteochondral defects in rabbit knee joints [11]. AAV vector is also used to deliver the gene encoding basic fibroblast growth factor (bFGF) into articular chondrocytes [71]. The transduced autologous cells are embedded into collagen gels and re-implanted into a full-thickness defect in the articular cartilage of the rabbit patellar groove. The transduction leads to the expression exceeding 8 weeks in >85 % of in vitro population and leads to the repair of articular cartilage defect [71]. AAV expressing receptor activator of nuclear factor κB ligand (RANKL) and vascular endothelial growth factor (VEGF) can also be freeze-dried into the allograft bone. Implantation of the coated allografts leads to marked formation of a new bone collar around the graft [72].

Currently, AAV appears to be the most promising vector for gene therapy, and may offer the best compromise between safety and efficacy for in vivo gene transfer [28, 65]. In 2012, Glybera®, an AAV vector designed to treat lipoprotein lipase deficiency, becomes the first gene therapy product approved in the Western world [73]. Additionally, an AAV vector expressing the TNF antagonist is employed in phase I and II clinical trials aiming for the treatment of rheumatoid arthritis (RA) [74]. The gene product is identical to etanercept (Enbrel®) used to treat RA patients and blocks the actions of TNF [75]. Intra-articular injection of this AAV vector exerts symptomatic benefit in some patients [74]. Although one subject dies in 2007 during the trial, the death is not attributed to AAV [76].

One challenge to the clinical application of AAV is that a large portion of human population possesses neutralizing antibodies against AAV [77], which diminishes the in vivo efficacy of AAV. Furthermore, the transduction efficiency of AAV vectors is hindered by the requirement to convert the singe stranded DNA genome into double stranded DNA prior to expression. This rate-limiting step prompts the development of self-complementary AAV (scAAV) vectors, which package an inverted repeat genome that can fold into double stranded DNA and can increase the transduction efficiency [78]. The trade-off of such scAAV is the loss of half of the cloning capacity.

Another challenge for the clinical application of AAV is the difficulties and high cost associated with production of high titer AAV, which requires transfection of producer cells with multiple plasmids. To overcome this problem, the genes required for AAV production can be cloned into separate baculovirus vectors (see Sect. 2.2.5), which, after co-infection of insect cells lead to the expression of AAV proteins and assembly of recombinant AAV vectors [79]. The new baculovirus/insect cell-based AAV production method is exploited for the production of Glybera®, the sole approved gene therapy product, and may encourage wider applications of AAV vectors in gene therapy.

### 2.2.5 *Baculovirus*

Baculoviruses are a diverse group of DNA viruses capable of infecting more than 500 insect species. Among the numerous baculoviruses, *Autographa californica* multiple nucleopolyhedrovirus (AcMNPV) contains a circular double-stranded DNA genome of $\approx$134 kb and is the most widely used. Budded AcMNPV is highly infectious to cultured insect cells, thus recombinant baculoviruses have been engineered to carry exogenous genes to infect insect cells for the production of numerous recombinant proteins (for review, see [80, 81]). Since the finding that baculovirus can efficiently transduce mammalian cells in the mid-1990s [82, 83], numerous permissive cells from different species have been discovered (for review, see [84–86]). Baculovirus neither replicates nor is toxic inside the transduced mammalian cells [84]. Baculoviral DNA degrades in the cells over time [87, 88] and there is no evidence of baculoviral DNA integration into host chromosomes unless selective pressure is applied [89]. These attributes minimize the potential side effects and ease the safety concerns. Furthermore, the large baculovirus genome confers a huge cloning capacity of at least 38 kb [90] and baculovirus can be propagated to high titers easily by infecting its natural host insect cells [86]. These properties have fueled growing interests to explore baculovirus for a wide variety of applications, ranging from protein production [91, 92], virus production [93–95] , virus-like particle production [88, 96, 97], eukaryotic protein display [98, 99], vaccine development [100–102], cancer therapy [103] to cell-based assay development [104–106].

Importantly, baculovirus is able to transduce primary chondrocytes derived from rats [87] and rabbits [107] with efficiencies exceeding 80 %. Baculovirus transduction does not hamper normal chondrocyte differentiation. Furthermore, baculovirus transduces bone marrow-derived mesenchymal stem cells (BMSCs) [108], adipose-derived stem cells (ASCs) [109] and even cell sheets derived from ASCs [110]. Under optimized conditions, the transduction efficiencies can exceed 95 %. Baculovirus also transduces adipogenic, chondrogenic and osteogenic progenitors originating from human BMSCs without obstructing the proliferation and differentiation potentials [111]. These properties spark the interests to develop baculovirus as a vector for the tissue engineering of bone, cartilage [112] and heart [110].

One shortcoming of baculovirus, however, is that baculovirus typically mediates transient (<7 days) transgene expression due to its non-replicating nature. Such transient expression may preclude the applications of baculovirus in certain scenarios requiring long-term sustained transgene expression (e.g. cancer therapy). To prolong the expression, attempts to incorporate AAV inverted terminal repeats [113, 114] or *Sleeping Beauty* transposon [115, 116] into baculovirus vectors have been made. For instance, a hybrid baculovirus exploiting the *Sleeping Beauty* transposon system is developed to extend the expression of microRNA [115] and anti-angiogenic factors for anti-cancer therapy [116].

Aside from these baculovirus vectors relying on transgene integration, hybrid baculovirus vectors enabling the episomal maintenance of transgene have been designed. We develop a hybrid dual baculovirus system in which one baculovirus expresses FLP recombinase while the substrate baculovirus harbors the transgene cassette flanked by two Frt sequences [108]. After co-transduction of mammalian cells with the two baculovirus vectors, the expressed FLP recognizes the Frt sites and excises the Frt-flanking cassette off the baculovirus genome, and hence catalyzes the recombination and formation of episomal DNA minicircles encompassing the transgene cassette. Such hybrid baculovirus vector successfully extends the transgene expression in a number of mammalian cells, including rabbit BMSCs [108] and ASCs [117]. The expression level and duration positively correlate with the recombination efficiency, presumably because the smaller DNA minicircle are less prone to nuclease attack and gene silencing [118, 119].

The excision/recombination efficiency is remarkably high, reaching 75 % in HEK293 cells, 85 % in BHK cells and 77 % in primary chondrocytes [108]. However, the FLP/Frt-mediated recombination efficiency occurs in only ≈40–50 % of rabbit BMSCs and ASCs [117, 120]. To further enhance the recombination efficiency, we have explored the codon-optimized FLP (FLPo), which can improve the FLP/Frt-mediated recombination at 37 °C [121]. Additionally, two other site-specific recombinases, Cre and codon-optimized ΦC31 (ΦC31o), have been tested. ΦC31o mediates excision/recombination between the heterotypic sites *attP* and *attB,* while Cre catalyzes excision/recombination events between two identical loxP sites [122]. Similar to the FLP/Frt-based baculovirus system, we construct a binary baculovirus vector system. Upon co-transduction, the transgene in the substrate baculovirus is excised by the recombinase (ΦC31o, Cre or FLPo) expressed by a second baculovirus vector and recombines into the smaller minicircle [123]. The recombination efficiency is lower by ΦC31o (≈40–75 %), but approaches ≈90–95 % by Cre and FLPo in various cell lines and stem cells such as human ASCs [123]. Compared with FLPo, Cre exerts higher expression level and lower cytotoxicity in human ASCs. The Cre/loxP-based baculovirus vectors are used to deliver genes encoding BMP-2 or VEGF into human ASCs, which results in efficient Cre/LoxP-mediated recombination and minicircle formation. As a result, the growth factor (BMP-2 or VEGF) expression is significantly prolonged and enhanced in human ASCs. The prolonged BMP2 expression ameliorates the osteogenesis of human ASCs, a stem cell with poor osteogenesis potential [123].

## References

1. Wang C-Y, Li F, Yang Y, Guo H-Y, Wu C-X, Wang S (2006) Recombinant baculovirus containing the Diphtheria toxin A gene for malignant glioma therapy. Cancer Res 66:5798–5806
2. Gao WT, Soloff AC, Lu XH, Montecalvo A, Nguyen DC, Matsuoka Y et al (2006) Protection of mice and poultry from lethal H5N1 avian influenza virus through adenovirus-based immunization. J Virol 80:1959–1964
3. Siltanen A, Kitabayashi K, Patila T, Ono M, Tikkanen I, Sawa Y et al (2011) Bcl-2 improves myoblast sheet therapy in rat chronic heart failure. Tissue Eng Part A 17:115–125
4. Siltanen A, Kitabayashi K, Lakkisto P, Makela J, Patila T, Ono M et al (2011) hHGF overexpression in myoblast sheets enhances their angiogenic potential in rat chronic heart failure. PLoS One 6:e19161

5. Rundle CH, Miyakoshi N, Kasukawa Y, Chen ST, Sheng MHC, Wergedal JE et al (2003) In vivo bone formation in fracture repair induced by direct retroviral-based gene therapy with bone morphogenetic protein-4. Bone 32:591–601
6. Evans CH, Ghivizzani SC, Herndon JH, Wasko MC, Reinecke J, Wehling P et al (2000) Clinical trials in the gene therapy of arthritis. Clin Orthop 379:S300–S307
7. Evans CH, Gouze JN, Gouze E, Robbins PD, Ghivizzani SC (2004) Osteoarthritis gene therapy. Gene Ther 11:379–389
8. Ghivizzani SC, Lechman ER, Tio C, Mule KM, Chada S, McCormack JE et al (1997) Direct retrovirus-mediated gene transfer to the synovium of the rabbit knee: implications for arthritis gene therapy. Gene Ther 4:977–982
9. Hsu WK, Sugiyama O, Park SH, Conduah A, Feeley BT, Liu NQ et al (2007) Lentiviral-mediated BMP-2 gene transfer enhances healing of segmental femoral defects in rats. Bone 40:931–938
10. Lieberman JR, Ghivizzani SC, Evans CH (2002) Gene transfer approaches to the healing of bone and cartilage. Mol Ther 6:141–147
11. Cucchiarini M, Madry H, Ma C, Thurn T, Zurakowski D, Menger MD et al (2005) Improved tissue repair in articular cartilage defects in vivo by rAAV-mediated overexpression of human fibroblast growth factor 2. Mol Ther 12:229–238
12. Gelse K, Schneider H (2006) Ex vivo gene therapy approaches to cartilage repair. Adv Drug Deliv Rev 58:259–284
13. Madry H, Kaul G, Cucchiarini M, Stein U, Zurakowski D, Remberger K et al (2005) Enhanced repair of articular cartilage defects in vivo by transplanted chondrocytes overexpressing insulin-like growth factor I (IGF-I). Gene Ther 12:1171–1179
14. Mierisch CM, Wilson HA, Turner MA, Milbrandt TA, Berthoux L, Hammarskjold ML et al (2003) Chondrocyte transplantation into articular cartilage defects with use of calcium alginate: the fate of the cells. J Bone Joint Surg Am 85A:1757–1767
15. Tsuchiya H, Kitoh H, Sugiura F, Ishiguro N (2003) Chondrogenesis enhanced by overexpression of sox9 gene in mouse bone marrow-derived mesenchymal stem cells. Biochem Biophys Res Commun 301:338–343
16. Madry H, Zurakowski D, Trippel SB (2001) Overexpression of human insulin-like growth factor-I promotes new tissue formation in an ex vivo model of articular chondrocyte transplantation. Gene Ther 8:1443–1449
17. Venkatesan N, Barre L, Benani A, Netter P, Magdalou J, Fournel-Gigleux S et al (2004) Stimulation of proteoglycan synthesis by glucuronosyltransferase-1 gene delivery: a strategy to promote cartilage repair. Proc Natl Acad Sci U S A 101:18087–18092
18. Kimelman-Bleich N, Pelled G, Zilberman Y, Kallai I, Mizrahi O, Tawackoli W et al (2011) Targeted gene-and-host progenitor cell therapy for nonunion bone fracture repair. Mol Ther 19:53–59
19. Wang YH, Ho ML, Chang JK, Chu HC, Lai SC, Wang GJ (2009) Microporation is a valuable transfection method for gene expression in human adipose tissue-derived stem cells. Mol Ther 17:302–308
20. Aslan H, Zilberman Y, Arbeli V, Sheyn D, Matan Y, Liebergall M et al (2006) Nucleofection-based ex vivo nonviral gene delivery to human stem cells as a platform for tissue regeneration. Tissue Eng 12:877–889
21. Dinser R, Kreppel F, Zaucke F, Blank C, Paulsson M, Kochanek S et al (2001) Comparison of long-term transgene expression after non-viral and adenoviral gene transfer into primary articular chondrocytes. Histochem Cell Biol 116:69–77
22. Park JS, Na K, Woo DG, Yang HN, Kim JM, Kim JH et al (2010) Non-viral gene delivery of DNA polyplexed with nanoparticles transfected into human mesenchymal stem cells. Biomaterials 31:124–132
23. Kim T-H, Kim M, Eltohamy M, Yun Y-R, Jang J-H, Kim H-W (2013) Efficacy of mesoporous silica nanoparticles in delivering BMP-2 plasmid DNA for in vitro osteogenic stimulation of mesenchymal stem cells. J Biomed Mater Res Part A 101A:1651–1660
24. Thomas CE, Ehrhardt A, Kay MA (2003) Progress and problems with the use of viral vectors for gene therapy. Nat Rev Genet 4:346–358

25. Somia N, Verma IM (2000) Gene therapy: trials and tribulations. Nat Rev Genet 1:91–99
26. Verma IM, Weitzman MD (2005) Gene therapy: twenty-first century medicine. Annu Rev Biochem 74:711–738
27. Yang F, Green JJ, Dinio T, Keung L, Cho SW, Park H et al (2009) Gene delivery to human adult and embryonic cell-derived stem cells using biodegradable nanoparticulate polymeric vectors. Gene Ther 16:533–546
28. Evans CH (2012) Gene delivery to bone. Adv Drug Deliv Rev 64:1331–1340
29. Lundstrom K (2003) Latest development in viral vectors for gene therapy. Trends Biotechnol 21:117–122
30. Airenne KJ, Hu Y-C, Kost TA, Smith RH, Kotin RM, Ono C et al (2013) Baculovirus: an insect-derived vector for diverse gene transfer applications. Mol Ther 21:739–749
31. Hu Y-C (2008) Baculoviral vectors for gene delivery: a review. Curr Gene Ther 8:54–65
32. Huang S, Kamihira M (2013) Development of hybrid viral vectors for gene therapy. Biotechnol Adv 31:208–223
33. Phillips JE, Gersbach CA, Garcia AJ (2007) Virus-based gene therapy strategies for bone regeneration. Biomaterials 28:211–229
34. Kofron MD, Laurencin CT (2006) Bone tissue engineering by gene delivery. Adv Drug Deliv Rev 58:555–576
35. Kimelman Bleich N, Kallai I, Lieberman JR, Schwarz EM, Pelled G, Gazit D (2012) Gene therapy approaches to regenerating bone. Adv Drug Deliv Rev 64:1320–1330
36. Blaese RM, Culver KW, Miller AD, Carter CS, Fleisher T, Clerici M et al (1995) T Lymphocyte-directed gene therapy for ADA – SCID: initial trial results after 4 years. Science 270:475–480
37. Hirschmann F, Verhoeyen E, Wirth D, Bauwens S, Hauser H, Rudert M (2002) Vital marking of articular chondrocytes by retroviral infection using green fluorescence protein. Osteoarthritis Cartilage 10:109–118
38. Yao SY, Sukonnik T, Kean T, Bharadwaj RR, Pasceri P, Ellis J (2004) Retrovirus silencing, variegation, extinction and memory are controlled by a dynamic interplay of multiple epigenetic modifications. Mol Ther 10:27–36
39. Bandara G, Mueller GM, Galealauri J, Tindal MH, Georgescu HI, Suchanek MK et al (1993) Intraarticular expression of biologically active interleukin-1 receptor antagonist protein by ex vivo gene transfer. Proc Natl Acad Sci U S A 90:10764–10768
40. Cavazzana-Calvo M, Hacein-Bey S, Basile CD, Gross F, Yvon E, Nusbaum P et al (2000) Gene therapy of human severe combined immunodeficiency (SCID)-X1 disease. Science 288:669–672
41. Hacein-Bey-Abina S, Von Kalle C, Schmidt M, McCcormack MP, Wulffraat N, Leboulch P et al (2003) LMO2-associated clonal T cell proliferation in two patients after gene therapy for SCID-X1. Science 302:415–419
42. Fischer A, Cavazzana-Calvo M (2008) Gene therapy of inherited diseases. Lancet 371:2044–2047
43. Hacein-Bey-Abina S, Garrigue A, Wang GP, Soulier J, Lim A, Morillon E et al (2008) Insertional oncogenesis in 4 patients after retrovirus-mediated gene therapy of SCID-X1. J Clin Invest 118:3132–3142
44. Hacein-Bey-Abina S, Hauer J, Lim A, Picard C, Wang GP, Berry CC et al (2010) Efficacy of gene therapy for X-linked severe combined immunodeficiency. New Eng J Med 363:355–364
45. Spitzer D, Hauser H, Wirth D (1999) Complement-protected amphotropic retroviruses from murine packaging cells. Hum Gene Ther 10:1893–1902
46. Gouze E, Pawliuk R, Pilapil C, Gouze JN, Fleet C, Palmer GD et al (2002) In vivo gene delivery to synovium by lentiviral vectors. Mol Ther 5:397–404
47. Li Y, Tew SR, Russell AM, Gonzalez KR, Hardingham TE, Hawkins RE (2004) Transduction of passaged human articular chondrocytes with adenoviral, retroviral, and lentiviral vectors and the effects of enhanced expression of SOX9. Tissue Eng 10:575–584
48. Yin GY, Liu WM, An P, Li P, Ding I, Planelles V et al (2002) Endostatin gene transfer inhibits joint angiogenesis and pannus formation in inflammatory arthritis. Mol Ther 5:547–554

49. Kato K, Miyake K, Igarashi T, Yoshino S, Shimada T (2005) Human immunodeficiency virus vector-mediated intra-articular expression of angiostatin inhibits progression of collagen-induced arthritis in mice. Rheumatol Int 25:522–529
50. Cartier N, Hacein-Bey-Abina S, Bartholomae CC, Veres G, Schmidt M, Kutschera I et al (2009) Hematopoietic stem cell gene therapy with a lentiviral vector in X-Linked adrenoleukodystrophy. Science 326:818–823
51. Ronen K, Negre O, Roth S, Colomb C, Malani N, Denaro M et al (2011) Distribution of lentiviral vector integration sites in mice following therapeutic gene transfer to treat beta-thalassemia. Mol Ther 19:1273–1286
52. Heckl D, Schwarzer A, Haemmerle R, Steinemann D, Rudolph C, Skawran B et al (2012) Lentiviral vector induced insertional haploinsufficiency of Ebf1 causes murine leukemia. Mol Ther 20:1187–1195
53. Schambach A, Zychlinski D, Ehrnstroem B, Baum C (2013) Biosafety features of lentiviral vectors. Hum Gene Ther 24:132–142
54. Banasik MB, McCray PB (2009) Integrase-defective lentiviral vectors: progress and applications. Gene Ther 17:150–157
55. Chick HE, Nowrouzi A, Fronza R, McDonald RA, Kane NM, Alba R et al (2012) Integrase-deficient lentiviral vectors mediate efficient gene transfer to human vascular smooth muscle cells with minimal genotoxic risk. Hum Gene Ther 23:1247–1257
56. Imperiale MJ, Kochanek S (2004) Adenovirus vectors: biology, design, and production. Curr Top Microbiol Immunol 273:335–357
57. Roessler BJ, Allen ED, Wilson JM, Hartman JW, Davidson BL (1993) Adenoviral-mediated gene transfer to rabbit synovium in vivo. J Clin Invest 92:1085–1092
58. Yao QP, Wang SJ, Glorioso JC, Evans CH, Robbins PD, Ghivizzani SC et al (2001) Gene transfer of p53 to arthritic joints stimulates synovial apoptosis and inhibits inflammation. Mol Ther 3:901–910
59. Ghivizzani SC, Lechman ER, Kang R, Tio C, Kolls J, Evans CH et al (1998) Direct adenovirus-mediated gene transfer of interleukin 1 and tumor necrosis factor alpha soluble receptors to rabbit knees with experimental arthritis has local and distal anti-arthritic effects. Proc Natl Acad Sci U S A 95:4613–4618
60. Brower-Toland BD, Saxer RA, Goodrich LR, Mi ZB, Robbins PD, Evans CH et al (2001) Direct adenovirus-mediated insulin-like growth factor I gene transfer enhances transplant chondrocyte function. Hum Gene Ther 12:117–129
61. Smith P, Shuler FD, Georgescu HI, Ghivizzani SC, Johnstone B, Niyibizi C et al (2000) Genetic enhancement of matrix synthesis by articular chondrocytes-Comparison of different growth factor genes in the presence and absence of interleukin-1. Arthritis Rheum 43:1156–1164
62. Nixon AJ, Haupt JL, Frisbie DD, Morisset SS, McIlwraith CW, Robbins PD et al (2005) Gene-mediated restoration of cartilage matrix by combination insulin-like factor-I/interleukin-1 receptor antagonist therapy. Gene Ther 12:177–186
63. Marshall E (1999) Gene therapy death prompts review of adenovirus vector. Science 286:2244–2245
64. Kochanek S, Clemens PR, Mitani K, Chen HH, Chan S, Caskey CT (1996) A new adenoviral vector: replacement of all viral coding sequences with 28 kb of DNA independently expressing both full-length dystrophin and beta-galactosidase. Proc Natl Acad Sci U S A 93:5731–5736
65. Evans CH, Gouze E, Gouze JN, Robbins PD, Ghivizzani SC (2006) Gene therapeutic approaches–transfer in vivo. Adv Drug Deliv Rev 58:243–258
66. Meager A (ed) (1999) Gene therapy technologies, applications and regulations. Wiley, New York
67. Mingozzi F, High KA (2011) Immune responses to AAV in clinical trials. Curr Gene Ther 11:321–330
68. Watanabe S, Imagawa T, Boivin GP, Gao GP, Wilson JM, Hirsch R (2000) Adeno-associated virus mediates long-term gene transfer and delivery of chondroprotective IL-4 to murine synovium. Mol Ther 2:147–152

69. Madry H, Cucchiarini M, Terwilliger EF, Trippel SB (2003) Recombinant adeno-associated virus vectors efficiently and persistently transduce chondrocytes in normal and osteoarthritic human articular cartilage. Hum Gene Ther 14:393–402
70. Ulrich-Vinther M, Stengaard C, Schwarz EM, Goldring MB, Soballe K (2005) Adeno-associated vector mediated gene transfer of transforming growth factor-β1 to normal and osteoarthritic human chondrocytes stimulates cartilage anabolism. Arthritis Rheum 46:2095–2104
71. Yokoo N, Saito T, Uesugi M, Kobayashi N, Xin K-Q, Okuda K et al (2005) Repair of articular cartilage defect by autologous transplantation of basic fibroblast growth factor gene-transduced chondrocytes with adeno-associated virus vector. Arthritis Rheum 52:164–170
72. Ito H, Koefoed M, Tiyapatanaputi P, Gromov K, Goater JJ, Carmouche J et al (2005) Remodeling of cortical bone allografts mediated by adherent rAAV-RANKL and VEGF gene therapy. Nat Med 11:291–297
73. Yla-Herttuala S (2012) Endgame: glybera finally recommended for approval as the first gene therapy drug in the European Union. Mol Ther 20:1831–1832
74. Mease PJ, Wei N, Fudman EJ, Kivitz AJ, Schechtman J, Trapp RG et al (2010) Safety, tolerability, and clinical outcomes after intraarticular injection of a recombinant adeno-associated vector containing a tumor necrosis factor antagonist gene: results of a phase 1/2 study. J Rheumatol 37:692–703
75. Evans CH, Ghivizzani SC, Robbins PD (2008) Arthritis gene therapy's first death. Arthritis Res Ther 10:110
76. Frank KM, Hogarth DK, Miller JL, Mandal S, Mease PJ, Samulski RJ et al (2009) Brief report: investigation of the cause of death in a gene-therapy trial. New Eng J Med 361:161–169
77. Nayak S, Herzog RW (2010) Progress and prospects: immune responses to viral vectors. Gene Ther 17:295–304
78. McCarty DM (2008) Self-complementary AAV, vectors; advances and applications. Mol Ther 16:1648–1656
79. Smith RH, Levy JR, Kotin RM (2009) A simplified baculovirus-AAV expression vector system coupled with one-step affinity purification yields high-titer rAAV stocks from insect cells. Mol Ther 17:1888–1896
80. Kost TA, Condreay JP (1999) Recombinant baculoviruses as expression vectors for insect and mammalian cells. Curr Opin Biotechnol 10:428–433
81. Kost TA, Condreay JP (2002) Recombinant baculoviruses as mammalian cell gene delivery vectors. Trends Biotechnol 20:173–180
82. Hofmann C, Sandig V, Jennings G, Rudolph M, Schlag P, Strauss M (1995) Efficient gene-transfer into human hepatocytes by baculovirus vectors. Proc Natl Acad Sci U S A 92:10099–10103
83. Boyce FM, Bucher NLR (1996) Baculovirus-mediated gene transfer into mammalian cells. Proc Natl Acad Sci U S A 93:2348–2352
84. Kost TA, Condreay JP, Jarvis DL (2005) Baculovirus as versatile vectors for protein expression in insect and mammalian cells. Nat Biotechnol 23:567–575
85. Hu Y-C (2005) Baculovirus as a highly efficient expression vector in insect and mammalian cells. Acta Pharmacol Sin 26:405–416
86. Hu Y-C (2006) Baculovirus vectors for gene therapy. Adv Virus Res 68:287–320
87. Ho Y-C, Chen H-C, Wang K-C, Hu Y-C (2004) Highly efficient baculovirus-mediated gene transfer into rat chondrocytes. Biotechnol Bioeng 88:643–651
88. Wang K-C, Wu J-C, Chung Y-C, Ho Y-C, Chang MD, Hu Y-C (2005) Baculovirus as a highly efficient gene delivery vector for the expression of hepatitis delta virus antigens in mammalian cells. Biotechnol Bioeng 89:464–473
89. Merrihew RV, Clay WC, Condreay JP, Witherspoon SM, Dallas WS, Kost TA (2001) Chromosomal integration of transduced recombinant baculovirus DNA in mammalian cells. J Virol 75:903–909
90. Cheshenko N, Krougliak N, Eisensmith RC, Krougliak VA (2001) A novel system for the production of fully deleted adenovirus vectors that does not require helper adenovirus. Gene Ther 8:846–854

91. Jardin BA, Zhao Y, Selvaraj M, Montes J, Tran R, Prakash S et al (2008) Expression of SEAP (secreted alkaline phosphatase) by baculovirus mediated transduction of HEK 293 cells in a hollow fiber bioreactor system. J Biotechnol 135:272–280
92. Liu CY-Y, Chen H-Z, Chao Y-C (2010) Maximizing baculovirus-mediated foreign proteins expression in mammalian cells. Curr Gene Ther 10:232–241
93. Lesch HP, Laitinen A, Peixoto C, Vicente T, Makkonen KE, Laitinen L et al (2011) Production and purification of lentiviral vectors generated in 293T suspension cells with baculoviral vectors. Gene Ther 18:531–538
94. Lesch HP, Turpeinen S, Niskanen EA, Mähönen AJ, Airenne KJ, Ylä-Herttuala S (2008) Generation of lentivirus vectors using recombinant baculoviruses. Gene Ther 15:1280–1286
95. Huang K-S, Lo W-H, Chung Y-C, Lai Y-K, Chen C-Y, Chou S-T et al (2007) Combination of baculovirus-mediated gene delivery and packed-bed reactor for scalable production of adeno-associated virus. Hum Gene Ther 18:1161–1170
96. Matsuo E, Tani H, Lim C, Komoda Y, Okamoto T, Miyamoto H et al (2006) Characterization of HCV-like particles produced in a human hepatoma cell line by a recombinant baculovirus. Biochem Biophys Res Commun 340:200–208
97. Chen Y-H, Wu J-C, Wang K-C, Chiang Y-W, Lai C-W, Chung Y-C et al (2005) Baculovirus-mediated production of HDV-like particles in BHK cells using a novel oscillating bioreactor. J Biotechnol 118:135–147
98. Ernst W, Schinko T, Spenger A, Oker-Blom C, Grabherr R (2006) Improving baculovirus transduction of mammalian cells by surface display of a RGD-motif. J Biotechnol 126:237–240
99. Grabherr R, Ernst W (2010) Baculovirus for eukaryotic protein display. Curr Gene Ther 10:195–200
100. Hu Y-C, Yao K, Wu T-Y (2008) Baculovirus as an expression and/or delivery vehicle for vaccine antigens. Expert Rev Vaccines 7:363–371
101. Madhan S, Prabakaran M, Kwang J (2010) Baculovirus as vaccine vectors. Curr Gene Ther 10:201–213
102. Tani H, Abe T, Matsunaga TM, Moiihi K, Matsuura Y (2008) Baculovirus vector for gene delivery and vaccine development. Future Virol 3:35–43
103. Wang S, Balasundaram G (2010) Potential cancer gene therapy by baculoviral transduction. Curr Gene Ther 10:214–225
104. Condreay JP, Ames RS, Hassan NJ, Kost TA, Merrihew RV, Mossakowska DE et al (2006) Baculoviruses and mammalian cell-based assays for drug screening. Adv Virus Res 68:255–286
105. Condreay JP, Kost TA (2007) Baculovirus expression vectors for insect and mammalian cells. Curr Drug Targets 8:1126–1131
106. Kost TA, Condreay JP, Ames RS (2010) Baculovirus gene delivery: a flexible assay development tool. Curr Gene Ther 10:168–173
107. Sung L-Y, Lo W-H, Chiu H-Y, Chen H-C, Chuang C-K, Lee H-P et al (2007) Modulation of chondrocyte phenotype via baculovirus-mediated growth factor expression. Biomaterials 28:3437–3447
108. Lo W-H, Hwang S-M, Chuang C-K, Chen C-Y, Hu Y-C (2009) Development of a hybrid baculoviral vector for sustained transgene expression. Mol Ther 17:658–666
109. Lu C-H, Lin K-J, Chiu H-Y, Chen C-Y, Yen T-C, Hwang S-M et al (2012) Improved chondrogenesis and engineered cartilage formation from TGF-β3-expressing adipose-derived stem cells cultured in the rotating-shaft bioreactor. Tissue Eng Part A 18:2114–2124
110. Yeh T-S, Dean Fang Y-H, Lu C-H, Chiu S-C, Yeh C-L, Yen T-C et al (2014) Baculovirus-transduced, VEGF-expressing adipose-derived stem cell sheet for the treatment of myocardium infarction. Biomaterials 35:174–184
111. Ho Y-C, Lee H-P, Hwang S-M, Lo W-H, Chen H-C, Chung C-K et al (2006) Baculovirus transduction of human mesenchymal stem cell-derived progenitor cells: variation of transgene expression with cellular differentiation states. Gene Ther 13:1471–1479

112. Lin C-Y, Lu C-H, Luo W-Y, Chang Y-H, Sung L-Y, Chiu H-Y et al (2010) Baculovirus as a gene delivery vector for cartilage and bone tissue engineering. Curr Gene Ther 10:242–254
113. Wang C-Y, Wang S (2005) Adeno-associated virus inverted terminal repeats improve neuronal transgene expression mediated by baculoviral vectors in rat brain. Hum Gene Ther 16:1219–1226
114. Zeng J, Du J, Zhao Y, Palanisamy N, Wang S (2007) Baculoviral vector-mediated transient and stable transgene expression in human embryonic stem cells. Stem Cells 25:1055–1061
115. Chen C-L, Luo W-Y, Lo W-H, Lin K-J, Sung L-Y, Shih Y-S et al (2011) Development of hybrid baculovirus vectors for artificial MicroRNA delivery and prolonged gene suppression. Biotechnol Bioeng 108:2958–2967
116. Luo WY, Shih YS, Hung CL, Lo KW, Chiang CS, Lo WH et al (2012) Development of the hybrid Sleeping Beauty-baculovirus vector for sustained gene expression and cancer therapy. Gene Ther 19:844–851
117. Lin C-Y, Lin K-J, Kao C-Y, Chen M-C, Yen T-Z, Lo W-H et al (2011) The role of adipose-derived stem cells engineered with the persistently expressing hybrid baculovirus in the healing of massive bone defects. Biomaterials 32:6505–6514
118. Lu J, Zhang F, Kay MA (2013) A mini-intronic plasmid (MIP): a novel robust transgene expression vector in vivo and in vitro. Mol Ther 21:954–963
119. Gracey Maniar LE, Maniar JM, Chen Z-Y, Lu J, Fire AZ, Kay MA (2013) Minicircle DNA vectors achieve sustained expression reflected by active chromatin and transcriptional level. Mol Ther 21:131–138
120. Lin C-Y, Chang Y-H, Kao C-Y, Lu C-H, Sung L-Y, Yen T-C et al (2012) Augmented healing of critical-size calvarial defects by baculovirus-engineered MSCs that persistently express growth factors. Biomaterials 33:3682–3692
121. Raymond CS, Soriano P (2007) High-efficiency FLP and PhiC31 site-specific recombination in mammalian cells. PLoS One 2:e162
122. Turan S, Galla M, Ernst E, Qiao JH, Voelkel C, Schiedlmeier B et al (2011) Recombinase-mediated cassette exchange (RMCE): traditional concepts and current challenges. J Mol Biol 407:193–221
123. Sung LY, Chen CL, Lin SY, Hwang SM, Lu CH, Li KC et al (2013) Enhanced and prolonged baculovirus-mediated expression by incorporating recombinase system and in cis elements: a comparative study. Nucleic Acids Res 41:e139

# Chapter 3
# Gene Therapy for Bone Tissue Engineering

**Abstract** Gene therapy has been employed in conjunction with bone engineering over the past decade, by which a variety of therapeutic genes are delivered to stimulate bone repair. These genes can be administered via in vivo or ex vivo approaches using either viral or nonviral vectors. This chapter reviews the fundamental aspects and recent progresses in the gene therapy-based bone engineering, with emphasis on the new genes, vectors and gene delivery approaches.

## 3.1 In Vivo Gene Delivery-Based Bone Engineering

For gene therapy-based bone engineering, the gene vector may be delivered in vivo by direct injection or by using gene activated matrix (GAM). Alternatively, the gene vector may be delivered ex vivo by using genetically modified cells (Fig. 3.1).

### *3.1.1 Direct Injection*

Direct gene vector injection intuitively provides the most straightforward and simplest method for gene delivery-based bone formation. In most studies involving direct injection, adenovirus is the most widely used vector, for the delivery of transgenes such as bone morphogenetic protein 2 (BMP-2), BMP-6, BMP-7 or BMP-9 (Table 3.1). The feasibility of in vivo adenovirus injection was established for ectopic bone formation in 1999 [17]. Evans and coworkers further demonstrate that

Y.-C. Hu, *Gene Therapy for Cartilage and Bone Tissue Engineering*,
SpringerBriefs in Bioengineering, DOI 10.1007/978-3-642-53923-7_3,

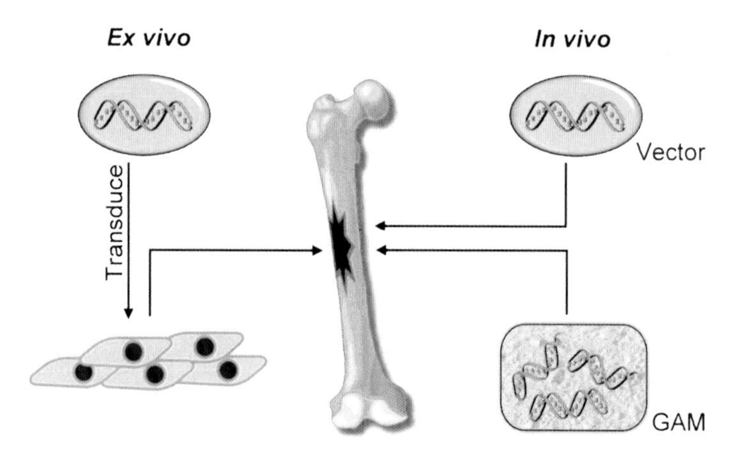


**Fig. 3.1 Different gene therapy methods for bone tissue engineering** (With permission of Lu et al. [1])

**Table 3.1** In vivo gene therapy for bone formation via direct injection (With permission of Lu et al. [1])

| Vector | Transgene | Model | References |
|---|---|---|---|
| Adenovirus | BMP-2 | Rabbit femoral segmental defect | [2] |
| Adenovirus | BMP-2 | Rat femoral segmental defect | [3–5] |
| Adenovirus | BMP-2, BMP-6 | Horse metatarsal | [6] |
| Adenovirus | BMP-2 | Sheep tibia | [7] |
| Adenovirus | BMP-6 | Rabbit ulna | [8] |
| Adenovirus | BMP-2, BMP-9 | Rat mandible | [9] |
| Adenovirus | BMP-4, BMP-6 | Nude rat muscle | [10] |
| Adenovirus | VEGF | Rabbit femur head necrosis | [11] |
| Retrovirus | BMP-2, BMP-4 | Rat femoral defect | [12] |
| Retrovirus | Cox-2 | Rat femoral fracture | [13] |
| Lentivirus | siRNA for HIF-1α and RunX2 | Rat Achilles tenotomy | [14] |
| AAV | BMP-2 | Mouse calvarial defect | [15] |
| Plasmid | BMP-9 | Mouse radial fracture | [16] |

*Cox-2* cycloxygenase-2

direct injection of an adenovirus expressing BMP-2 successfully triggers the ossification of segmental bone defects at the femora of rabbits [2] and rats [3–5]. Bone healing efficiency is pronouncedly affected by the vector dose and injection timing, as higher vector dose [4, 5] and delayed administration of such adenovirus (AdBMP2) 5 days after surgery improves the repair. However, the newly formed bone lacks the structural organization and mechanical strength of native bone [3, 4]. Bone healing is also observed in horses injected with AdBMP2 or AdBMP6 [6] but not in sheep injected with AdBMP2 [7]. In addition, cycloxygenase-2 (Cox-2) is an enzyme that facilitates the production of prostaglandins that promote angiogenesis and bone formation. Bony union of the femoral fracture can be achieved by direct injection of a retrovirus expressing Cox-2 into the rat femoral fracture [13]. Furthermore, direct injection of an adenovirus expressing VEGF into the femur head necrotic regions of rabbits promotes bone formation and re-vascularization in the subchondral necrotic region of the femoral head, thus indirectly protecting the necrotic bone trabecula from adsorption [11].

In contrast to anabolic growth factors delivered by adenovirus, lentiviral vectors that express small interfering RNA (siRNA) against hypoxia-inducible factor 1α (HIF-1α) and Runt-related transcription factor 2 (RunX2) [14] are developed. HIF is a transcription factor that directly increases VEGF gene expression and is associated with coupled regulation of angiogenesis and osteogenesis [18]. RunX2 is a key transcription factor associated with osteoblast differentiation. To prevent heterotopic ossification that often results from traumatic injury, lentiviral vectors encoding HIF-1α-siRNA and RunX2-siRNA are injected into rats that undergo Achilles tenotomy, and lentivirus-mediated inhibition of HIF-1α and RunX2 is able to inhibit heterotropic ossification formation [14].

Aside from viral vectors, direct injection of plasmid encoding BMP-7 is possible [19], but bone formation is very poor presumably due to low transfection efficiency. To enhance the transfection efficiency, Gazit and coworkers have developed an in vivo electroporation method [16]. Ten days after creating nonunion fractures in the radii of mice and implantation of a collagen sponge, plasmid DNA encoding BMP-9 is injected into the radial defect site, followed by electroporation using needle electrodes placed at both sides of the radial defect (1–2 mm apart). The in vivo electroporation results in bone bridging and improves bone formation [16].

One hurdle to the direct injection is the potential spreading of the vectors to non-target sites, which otherwise may elicit heterotopic ossification of adjacent muscle tissue and fusion of one bone to an adjacent bone [20]. If the injection site is located near a joint compartment, treatment may induce ossification of cartilaginous and ligamentous tissues, leading to joint dysfunction [21].

### *3.1.2 Gene Activated Matrix (GAM)*

The gene can also be delivered via gene activated matrix (GAM), which enables controlled, slow release of the gene vector to the surrounding cell or tissue. The materials used as the GAM include collagen, silk, chitosan, polymer, composite material, demineralized bone or even allograft bone (Table 3.2). Again, adenovirus and plasmid are the most commonly used vectors embedded within the GAM for delivery, and BMPs and vascular endothelial growth factor (VEGF) are the growth factors most often employed. Alternatively, Nell-1, a novel osteoinductive gene [44], is delivered via adenovirus within the demineralized bone matrix (DBM). Implantation of the DBM carrier containing the adenovirus into athymic rats improves the spinal fusion [37]. Besides, platelet-derived growth factor B (PDGF-B) is a mitogen for fibroblasts and a cytokine capable of recruiting mesenchymal cells to sites of injury [45]. A GAM comprising mesoporous bioglass/silk fibrin scaffold

**Table 3.2** Types of GAM used for bone formation (Adapted from Lu et al. [1])

| GAM | Vector | Transgene | Models | References |
|---|---|---|---|---|
| Collagen | Plasmid | BMP-4, PTH 1-34 | Rat femoral defect | [22] |
| Collagen | Plasmid | PTH 1-34 | Dog tibial defect; horse metacarpal | [23, 24] |
| Collagen | Plasmid | VEGF | Rabbit radial defect | [25] |
| Collagen | Plasmid/calcium phosphate | BMP-2 | Rat tibial defect | [26] |
| Collagen/calcium phosphate | Plasmid | VEGF | Mouse femoral defect | [27] |
| Polyplex nanomicelle | Plasmid | caALK6, RunX2 | Mouse cranial defect | [28] |
| PLGA | Plasmid condensed with PEI | BMP-4 | Rat cranial defect | [29] |
| Triacrylate/amine-gelatin | Plasmid | BMP-2 | Rat cranial defect | [30] |
| Fibronectin/apatite | Plasmid | BMP-2 | Rat cranial defect | [31, 32] |
| Silk fibroin | Adenovirus | BMP-7 | Mouse cranial defect | [33] |
| Silk fibrin/bioglass | Adenovirus | BMP-7, PDGF-B | Rat femoral defect | [34] |
| Chitosan/collagen | Adenovirus | BMP-7, PDGF-B | Dog dental implant | [35] |
| Chitosan/collagen | Adenovirus | BMP-2, VEGF | Dog dental implant | [36] |
| DBM | Adenovirus | Nell-1 | Rat spinal fusion | [37] |
| Muscle or fat tissue | Adenovirus | BMP-2 | Rat femoral defect | [38] |
| Muscle tissue | Adenovirus | BMP-2 | Rat calvarial defect | [39] |
| Bone allograft | AAV | VEGF, RANKL | Mouse femoral defect | [40] |
| Bone allograft | AAV | caALK2 | Mouse femoral defect | [41] |
| Bone allograft | AAV | BMP-2 | Mouse femoral defect | [42] |
| Bone allograft | AAV | BMP-2 | Mouse calvarial defect | [43] |

*PTH* parathyroid hormone, *PDGF-B* platelet-derived growth factor B, *DBM* demineralized bone, *caALK6* constitutively active form of activin receptor-like kinase 6, *RANKL* receptor activator of nuclear factor κB ligand

and adenovirus expressing PDGF-B and adenovirus expressing BMP-7 is recently fabricated for bone healing in osteoporotic rats. Implantation of this GAM into the critical-size femoral defect in ovariectomised rats leads to new bone formation and initiation of bone turnover and remodeling [34].

Moreover, structural allograft healing is often limited because of a lack of vascularization and remodeling. It has been uncovered that allografts are deficient in the expression of VEGF and receptor activator of nuclear factor κB ligand (RANKL) [40], which are known to dominantly regulate angiogenesis and osteoclastic bone resorption. In light of the importance of RANKL and VEGF, Ito et al. freeze dry AAV vectors encoding RANKEL and VEGF onto the cortical surface of allograft without losing infectivity [40]. Implantation of the AAV-RANKL- and AAV-VEGF-coated allografts into a mouse model leads to marked remodeling and vascularization, hence stimulating the formation of a new bone collar around the graft and revitalizing the structural allografts [40].

In a more recent study, Yazici et al. use an allograft coated with a new self-complementary AAV expressing BMP-2 (scAAV2.5-BMP2) to repair the segmental bone at the femora. After 6 weeks, the AAV-coated allografts form a new cortical shell that resembles live allografts and revitalization of the allograft is observed, as evidenced by the live bone marrow within and around the necrotic cortical bone [42].

## 3.2 Ex Vivo Gene Delivery-Based Bone Formation

### *3.2.1 Systemic Delivery*

For ex vivo gene therapy, the genetically modified cells may be injected/infused for systemic delivery and dissemination, which is particularly useful for applications such as osteoporosis or osteogenesis imperfecta [46]. For systemic delivery, cells capable of homing to bone and differentiating into osteoblasts are particularly ideal. Therefore, mesenchymal stem cells (MSCs) have captured attention as they can home to injury site and differentiate into osteoblasts [47]. Indeed, in a mouse tibia fracture model, intravenous injection of bone marrow-derived MSCs (BMSCs) leads to cell migration to the fracture site, cell engraftment within the callus endosteal niche and improvement of the fracture healing [48]. Furthermore, intravenous injection of mouse BMSCs transduced with AAV6 encoding BMP-2/VEGF enhances bone formation and vascularity in a nude mouse model of segmental tibial defect [49]. However, most of the injected BMSCs are trapped in the lungs and liver, despite being able to home to the tibia defect site. Such poor homing efficiency to bone agrees with the observation that ≈98 % of the BMSCs are lost to the liver and spleen after intravenous injection [50]. The injected BMSCs do not integrate into the newly formed bone [49] but exert bone healing effects through the so called "touch and go" effects [51], as BMSCs express high levels of various growth factors involved in the repair process [49].

The MSCs homing to bone is dependent on CXCR4 (CXC chemokine receptor-4) [47, 48], thus Lien et al. have attempted to improve homing by engineering BMSCs with an adenovirus encoding CXCR-4. Injection of BMSCs co-transduced with adenovirus expressing CXCR-4 and RunX2 restores the bone mass and mechanical strength in an osteoporotic mouse model [52]. Similarly, intravenous injection of retrovirus-engineered BMSCs capable of over-expressing receptor activator of nuclear factor-κB (RANK)-Fc and CXCR4 promotes the in vivo cell trafficking to bone in ovariectomy-induced osteoporotic mice and prevents bone loss [53]. Interestingly, a recent study shows that intravenous injection of peptidomimetic ligand (LLP2A)-bisphosphonate (alendronate, Ale) enhances the recruitment of BMSCs to the bone surface, and improves the bone formation and bone strength [54]. Thus LLP2A-Ale may be co-injected with the genetically engineered BMSCs to synergize the bone healing effect. Alternative to BMSCs, Hall et al. use stem cell antigen-1-positive (Sca-1+) hematopoietic cells that are transduced with a retrovirus expressing fibroblast growth factor-2 (FGF-2). Retro-orbital injection of the engineered Sca-1+ cells into mice results in long-term engraftment, higher serum FGF-2 level and massive endosteal bone formation [55].

### *3.2.2 Local Delivery*

Despite the promise of systemic delivery, most of the literature describes local delivery of cells genetically modified ex vivo to accelerate the healing of fractures or segmental bony defects. Early studies have demonstrated the proof-of-concept of such cell-based gene therapy in ectopic bone formation (e.g. implantation into muscles), which have been reviewed elsewhere [56–58], thus this section focuses on orthotopic bone formation/regeneration (Table 3.3). Most of these studies have exploited growth factors in the BMP family and osteoprogenitor cells such as BMSCs, adipose-derived stem cells (ASCs) or muscle-derived stem cells (MDSCs).

Since natural bone fracture healing process requires coordinated coupling between osteogenesis and angiogenesis [97], Huard and coworkers have genetically engineered MDSCs to express BMP-4 or VEGF by retrovirus, mixed the cells at selected ratios, and implanted the cells into the critical-size calvarial defects in mice [69]. They demonstrate that VEGF acts synergistically with BMP-4 to enhance the calvarial bone formation via the endochondral ossification pathway, by increasing the recruitment of MSCs, enhancing cell survival and augmenting cartilage formation in the early stages of endochondral bone formation [69]. A subsequent study also shows that simultaneous expression of BMP-2 and VEGF by retrovirus-transduced MDSCs stimulates calvarial bone formation [71]. The supportive roles of VEGF on calvarial bone healing mediated by BMP-2 are further confirmed later [98, 99]. The osteoinductive and angiogenic effects of BMP-2/VEGF have prompted the combined use of both factors in recent years to synergistically promote the healing of cranial [99], ulnar [100] and femoral [101] bone defects.

Despite the synergistic bone healing effects imparted by BMP-2 and VEGF, it should be noted, however, that the ratio between VEGF and BMP-2 influences their synergistic interaction, with a higher proportion of VEGF leading to decreased synergism [71]. Concurrent with this finding, implantation of retrovirus-engineered

**Table 3.3** Selected examples of ex vivo gene therapy for orthotopic bone repair (Adapted from Lu et al. [1])

| Vector | Cells | Transgene | Models | References |
|---|---|---|---|---|
| Plasmid | BMSCs | BMP-2 | Mouse tibial defect | [59] |
| Plasmid | BMSCs | BMP-2 | Mouse spinal fusion | [60] |
| Plasmid[a] | ASCs | BMP-6 | Mouse spinal fusion | [61] |
| Plasmid[a] | ASCs | BMP-6 | Rat vertebral bone void defect | [62] |
| Retrovirus | BMSCs, ASCs | Sonic hedgehog | Rabbit calvarial defect | [63] |
| Retrovirus | BMSCs | RunX2 | Rat calvarial defects | [64] |
| Retrovirus | BMSCs | BMP-4 | Rat calvarial defect | [65] |
| Retrovirus | MDSCs | BMP-2 | Mouse calvarial defect | [66] |
| Retrovirus | MDSCs | BMP-4 | Mouse calvarial defect | [67] |
| Retrovirus | MDSCs | BMP-4 | Rat calvarial defect | [68] |
| Retrovirus | MDSCs | BMP-4, VEGF | Mouse calvarial defect | [69] |
| Retrovirus | MDSCs | BMP-4, noggin | Mouse calvarial defect | [70] |
| Retrovirus | MDSCs | BMP-2, VEGF | Mouse calvarial defect | [71] |
| Retrovirus | iPSCs | SATB2 | Mouse calvarial defect | [72] |
| Adenovirus | BMSCs | BMP-2 | Mouse radius defect | [73] |
| Adenovirus | BMSCs | BMP-2 | Rat calvarial defect | [74] |
| Adenovirus | BMSCs | BMP-2 | Goat tibial defect | [75] |
| Adenovirus | ASCs | BMP-2 | Rat femoral defect | [76] |
| Adenovirus/plasmid | BMSCs | BMP-2 | Rat mandibular defect | [77] |
| Adenovirus | Bone marrow cells | BMP-2 | Rat femoral defects | [78] |
| Adenovirus | BMSCs | BMP-2 | Rat femoral defect | [79] |
| Adenovirus | BMSCs | BMP-2 | Rat mandibular defect | [80] |
| Adenovirus | BMSCs | RunX2 | Mouse calvarial defect | [81] |

(continued)

**Table 3.3** (continued)

| Vector | Cells | Transgene | Models | References |
|---|---|---|---|---|
| Adenovirus | Skin fibroblasts | LMP 3 | Rat mandibular defect | [82] |
| Adenovirus | BMSCs and EPC | BMP-2 | Rat calvarial defect | [83] |
| Lentivirus/ Adenovirus | Bone marrow cells | BMP-2 | Mouse radial defect | [84] |
| Lentivirus/ Adenovirus | Bone marrow cells | BMP-2 | Rat femoral defect | [85] |
| Lentivirus | BMSCs | BMP-2 | Rat femoral defect | [86] |
| Lentivirus | Bone marrow buffy coat cells | BMP-2 | Rat femoral defect | [87] |
| Lentivirus | BMSCs | HIF-1α | Rat calvarial defect | [88] [89] [18] |
| Lentivirus | ASCs | miR-31, miR-31 anti-sense | Rat calvarial defect | [90] |
| miRNA mimics (agomer) | BMSCs | miR-26a | Mouse calvarial defect | [91] |
| Baculovirus | BMSCs | BMP-2 | Rat calvarial defect | [92] |
| Baculovirus | BMSCs | BMP-2, VEGF | Rabbit femoral defects | [93] |
| Baculovirus | ASCs | BMP-2, VEGF | Rabbit femoral defects | [94, 95] |
| Baculovirus | BMSCs | BMP-2, VEGF | Rabbit calvarial defect | [96] |
| Baculovirus | ASCs | BMP-2, miR-148b | Mouse calvarial defect | Unpublished data |

[a]Plasmid delivered by nucleofection. *SATB2* special AT-rich sequence-binding protein 2, *LMP 3* Lim mineralization protein 3, *miR-26a* miRNA-26a, *miR-148b* miRNA-148b

C2C12 or NIH/3T3 cells that express BMP-4/VEGF into the muscle pocket in SCID mice leads to ectopic bone formation [102], but the cells expressing both BMP-4 and VEGF display significantly less bone formation than the same cells expressing only BMP-4. The ectopic bone formation is impaired when the ratio of VEGF to BMP-4 is high, but the detrimental effect on bone formation disappears when the ratio is low. Therefore, the VEGF's synergistic effect on BMP-4-induced ectopic bone formation is dose and cell-type dependent [102].

Recently, Helmrich et al. generate osteogenic grafts with an increased vascularization potential in an ectopic nude rat model in vivo, by genetically modifying human BMSCs with retrovirus to express rat VEGF [103]. The transduced BMSCs are loaded onto silicate-substituted apatite granules and implanted. Eight weeks after implantation, the VEGF-expressing BMSCs significantly increase the vascular density in the grafts, consisting of physiologically structured vascular networks with both conductance vessels and capillaries. However, VEGF specifically causes a global reduction in bone quantity as manifested by thin trabeculae of immature matrix. VEGF does not impair BMSCs engraftment in vivo, but strongly increases the recruitment of TRAP- and Cathepsin K-positive osteoclasts [103]. These data suggest that VEGF overexpression effectively improves the vascularization of osteogenic grafts but concomitantly promotes bone resorption, which might explain why the ratio of BMP and VEGF is important.

Besides VEGF, HIF is a transcription factor that directly increases VEGF gene expression and is a major regulator of angiogenic-osteogenic coupling, which prompts the employment of lentiviral vectors that express HIF-1α for genetic modification of rat BMSCs [18]. The transduced cells are seeded to scaffolds and implanted to 5 mm critical-size calvarial defects in Fisher 344 rats. The HIF-1α-overexpressing BMSCs remarkably improve the repair of calvarial defects in rats, as manifested by the increased bone volume, bone mineral density, blood vessel number/area and blood flow. HIF-1α-overexpression in BMSCs significantly enhances the expression of key angiogenic factors including VEGF and stromal-derived factor (SDF-1) at both mRNA and protein levels [89]. As a result, the HIF-1α-overexpressing BMSCs dramatically improve blood vessel formation in the tissue-engineered bone [18, 88].

Alternatively, angiogenesis may be stimulated by co-implantation with endothelial cells (EC) or endothelial progenitor cells (EPC). Implantation of 3D poly (L-lactide-*co*-glycolide) (PLGA) sintered microsphere scaffolds containing EC and AdVEGF-transduced ASCs into mice results in marked vascular growth within the PLGA scaffolds [104]. Further, He et al. transduce BMSCs and EPCs with AdBMP2 and implant the cells together with an injectable porous nano calcium sulfate/alginate (nCS/A) scaffold into the rat critical-size calvarial bone defect [83]. Combination of BMP-2 gene-modified BMSCs and EPCs in nCS/A substantially increases the new bone and vascular formation. In particular, the EPCs ameliorate new vascular growth, and BMP-2 gene modification of BMSCs and EPCs remarkably augments bone regeneration.

As mentioned in Chap. 2, baculovirus has emerged as a promising vector for stem cell modification and bone tissue engineering [105, 106]. Baculovirus can effectively transduce human BMSCs [107] and the transduction efficiencies are comparable or superior to those obtained by retroviral or adenoviral vectors [108, 109] and can be further elevated to >95 % under optimized conditions [110]. Therefore, we genetically engineer human BMSCs with a BMP-2-expressing baculovirus (Bac-CB) and confirm that Bac-CB transduction directs in vitro commitment of naïve BMSCs into osteoblasts in a virus dose-dependent manner [111]. The BMP-2 expression level is remarkably lower in the human BMSCs (<10 ng/ml) than in the rabbit articular chondrocytes (≈500–1,000 ng/ml), indicating that BMP-2 expression using the same baculovirus varies with cell type. Despite the transient and lower BMP-2 expression level, Bac-CB transduction at a multiplicity of infection (MOI) 40 induces the BMSCs to differentiate into late osteoblast stage as evidenced by ≈3–4 fold stimulation of alkaline phosphatase (ALP) expression at day 9 and declining ALP expression thereafter [111]. Bac-CB supertransduction at day 6 further accelerates the differentiation progression as judged by the calcium deposition stained by Alizarin red and reverse transcription PCR (RT-PCR) analysis of osteopontin and osteocalcin [111]. To explore the ectopic bone formation, the transduced BMSCs are suspended in the alginate solution and co-injected with $CaCl_2$ solution into the back subcutis of immunodeficient nude mice, which results in immediate encapsulation in situ. Two weeks after implantation, no matrix accumulates in the animals implanted with mock-transduced BMSCs, indicating no spontaneous osteogenesis in the subcutis. However, the Bac-CB-engineered BMSCs give rise to dense deposition of calcium and osteocalcin in the matrix, progressive mineralization and ectopic bone formation [111].

Whether the baculovirus-engineered human BMSCs are tolerant in immunocompetent animals and heal the critical-sized calvarial bone defect is investigated in a subsequent study [92]. BMSCs are transduced by Bac-CB as described earlier [111], followed by cell seeding to PLGA scaffolds and transplantation into the critical-sized defects (8 mm in diameter) at the rat calvaria. Without immunosuppression, Bac-CB transduction substantially boosts the BMSCs aggregation (which signals the onset of calvarial bone formation), ameliorates the accumulation of mineralized bone matrix and initiates the bone island formation at week 4. However, the xenogeneic human BMSCs undergo rejection responses as evidenced by the infiltration of macrophages, $CD3^+$ and $CD8^+$ T cells into the graft, as well as the eradication of transplanted donor cells at week 12. With the administration of immunosuppressive drugs, Bac-CB-engineered human BMSCs enhance the trabecular bone formation at week 12 and prolong cell survival [92]. However, the xenotransplanted BMSCs are eventually rejected even with immunosuppression, filling only $\approx 28\ \%$ of the original defect area. The incomplete bone healing is presumably ascribed to the augmented osteogenic differentiation of BMSCs that leads to the loss of their immunoprivileged properties [112]. These data altogether show that Bac-CB holds promise for BMSCs engineering and calvarial bone repair, but the use of human BMSCs cannot overcome the immunological barrier in xenogeneic recipients.

To circumvent the acute rejection resulting from xenotransplantation, in a more recent study the BMSCs isolated from New Zealand White (NZW) rabbits are used for baculovirus transduction and allotransplantation [93]. Given the roles of VEGF in angiogenesis and ossification, a recombinant baculovirus (Bac-CV) expressing VEGF is constructed [93]. After in vitro transduction, the BMSCs transduced with Bac-CV and Bac-CB are mixed at a number ratio of 1:4 and seeded into PLGA scaffolds. The cell/scaffold constructs are implanted to the critical-sized femoral segmental defects of allogeneic, immunocompetent NZW rabbits [93]. The constructs not only accelerate the bone healing (bridging of the defects occurs in all 13 animals at as early as week 4) compared with the controls, but also give rise to conspicuous formation of trabecular and cortical bones as well as new blood vessels at week 8. As a result, the torsional stiffness of the healed femora approaches $\approx 90\ \%$ of the uninjured bones. These data concretely confirm that BMSCs engineered by baculoviruses expressing BMP-2 and VEGF synergistically augment the healing of large femoral segmental bone defects in immunocompetent animals. The synergism is attributable to the improved angiogenesis [93], which may enhance the cell survival and facilitate the migration of host osteoprogenitor cells to the bone regeneration site [69].

Aside from BMSCs, ASCs have gained growing popularity for bone regeneration because ASCs can be easily isolated from liposuction in large quantities, but ASCs are suggested to be inferior to BMSCs in osteogenesis potential [113, 114]. Indeed, ASCs engineered by the baculovirus vectors transiently expressing BMP-2/VEGF (Bac-CB and Bac-CB, denoted as S group) lead to poor healing of segmental femoral bone defects (Fig. 3.2). To use ASCs for repairing large, segmental bone defects, we surmise that sustained expression of factors promoting osteogenesis (BMP-2) and angiogenesis (VEGF) is necessary. As such, we have developed the hybrid baculovirus system based on FLP/Frt-mediated recombination and DNA minicircle formation (see Chap. 2). The FLP/Frt-mediated recombination occurs in the NZW rabbit ASCs, enabling persistent transgene expression for >28 days [94]. Allotransplantation of the NZW rabbit ASCs transduced with the hybrid baculoviruses expressing BMP-2/VEGF (designated as L group, Fig. 3.2) into the critical-size femoral segmental defects accelerates the healing, improves the bone quality and angiogenesis when compared with the S group (Fig. 3.2).

The progression of bone remodeling gives rise to the resorption of trabecular bone, conspicuous reconstruction of medullary cavity and cortical bone with lamellar structure at 8 months post-transplantation, hence conferring mechanical properties that are comparable to those of non-operated femora (unpublished data). Therefore, the hybrid baculovirus-engineered ASCs and prolonged BMP-2/VEGF expression not only heal and remodel the massive segmental defects, but also revitalize the defects into living bone tissues that structurally and biomechanically resemble intact bones.

The hybrid baculovirus vectors are also used to transduce BMSCs and the prolonged BMP-2/VEGF expression promotes the healing of critical-size (8 mm) calvarial defects in rabbits [96]. However, when ASCs are used as the cell source, calvarial bone repair is barely observed [96]. In nature, calvarial bone forms via intramembranous ossification without cartilage templates. However, it is suggested that chondrocytes/cartilages promote calvarial healing [97]. Moreover, Scotti et al. recently induce BMSCs to various chondrogenic differentiation stages in vitro and subcutaneously implant the chondrogenitor cells into nude mice [115]. Bone trabeculae formation occurs only when BMSCs have pre-differentiated into hypertrophic tissue structures, and advanced chondrogenic maturation in vitro accelerates the formation of larger bony tissues in vivo [115]. These studies suggest that inducing the endochondral ossification pathway may stimulate in vivo bone formation. Therefore, we hypothesize that inducing ASCs chondrogenesis and endochondral ossification involving cartilage formation can improve calvarial healing. To evaluate this hypothesis and selectively induce osteogenesis/chondrogenesis, rabbit ASCs are engineered by baculovirus vector expressing either osteoinductive (BMP-2) or chondroinductive (transforming growth factor β3 (TGF-β3)) factor [116]. The transduced ASCs are seeded into either apatite-coated PLGA (which preferentially induces osteogenesis) or gelatin sponge (which preferentially

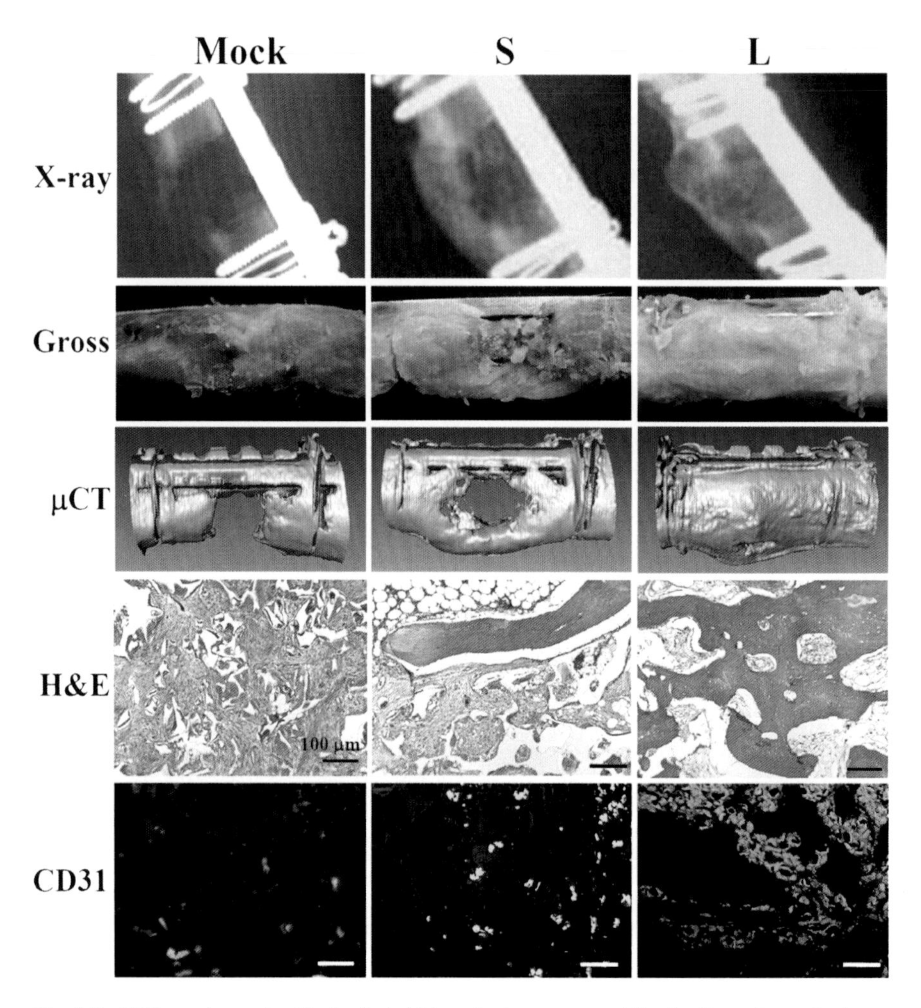


**Fig. 3.2 ASCs engineered with the hybrid baculovirus augment the healing of massive bone defects**. The NZW rabbit ASCs are transduced with the hybrid baculovirus vectors conferring sustained expression of BMP-2 or VEGF, mixed at a number ratio of 4:1, loaded into cylindrical PLGA scaffolds and implanted to the critical-sized segmental defects at the femora of NZW rabbits (designated L group). The S group comprises ASCs that are transduced with conventional baculoviruses transiently expressing BMP-2/VEGF and implanted in a similar fashion. The Mock group contains the mock-transduced ASCs as the negative control. X-ray radiography, gross appearance examination, μCT analyses, hematoxylin & eosin (H&E) staining and CD31-specific immunohistochemical staining (to detect blood vessel formation) performed at 12 weeks post-implantation collectively demonstrate that the L group results in significantly improved bone healing and angiogenesis in comparison with the S and Mock groups (With permission of Lin et al. [94])

promotes chondrogenesis) scaffolds, and allotransplanted into critical-size calvarial defects [116]. Among the 4 ASCs/scaffold constructs, gelatin constructs elicit in vitro chondrogenesis, in vivo osteogenic metabolism and calvarial healing more effectively than apatite-coated PLGA, regardless of BMP-2 or TGF-β3 expression

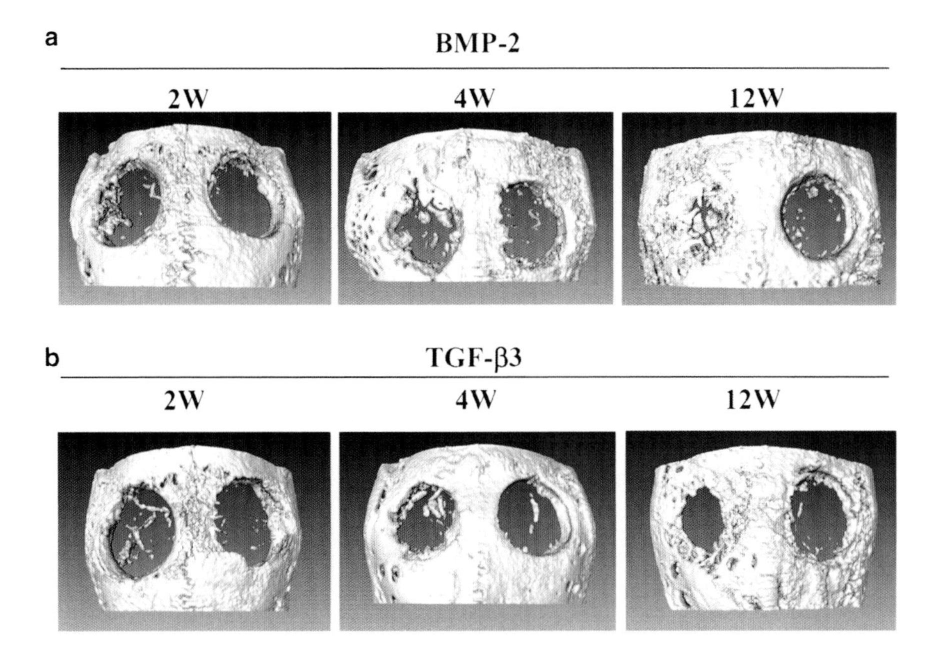


**Fig. 3.3 Effects of growth factor expression and scaffold material on calvarial bone healing.** (**a**) Effects of BMP-2. (**b**) Effects of TGF-b3. The NZW rabbit ASCs are transduced with the new codon-optimized FLPo/Frt-based hybrid baculovirus vector expressing BMP-2 or TGF-β3, and seeded into either apatite-coated PLGA or gelatin sponges. For head-to-head comparison of scaffolds, two defects (8 mm in diameter) are created on the right and left parietal bones in NZW rabbits and ASCs/PLGA and ASCs/gelatin constructs are implanted to the right and left defects, respectively. The calvarial bone repair is evaluated by μCT analysis of the skulls removed at week 2 (2W), 4 (4W) and 12 (12W). In the BMP-2 group, the 3D rendering images confirm progressive and substantial bone healing on the gelatin side (left side) from 2W to 12W but the bone repair is slow and poor on the PLGA side (right side). In the TGF-β3 group, similarly the gelatin constructs trigger better bone healing than the PLGA constructs at 2W, 4W and 12W, confirming that gelatin constructs provoke more effective calvarial bone healing. However, at 12W the gelatin constructs result in bone healing that is inferior to the repair induced by the gelatin constructs in the BMP-2 group, attesting that BMP-2 is more effective than TGF-β3 for calvarial bone healing (With permission of Lin et al. [116])

(Fig. 3.3). The BMP-2-expressing ASCs/gelatin constructs trigger better bone healing than TGF-β3-expressing ASCs/gelatin, filling ≈86 % of the defect area and ≈61 % of the volume at week 12, indicating that BMP-2 is more effective than TGF-β3 for calvarial bone healing [116]. Such healing mediated by the BMP-2-expressing ASCs/gelatin constructs is dramatically improved when compared with the healing using ASCs expressing BMP-2/VEGF [96]. The healing proceeds via endochondral ossification, instead of intramembranous pathway, as evidenced by the formation of cartilage that undergoes osteogenesis and hypertrophy. These data demonstrate that the BMP-2-expressing ASCs/gelatin constructs are able to switch the ossification pathway and significantly augment calvarial healing. This study also underscores the importance of growth factor/scaffold combinations on the healing efficacy and pathway [116].

Alternatively, calvarial bone healing can be achieved by baculovirus-mediated microRNA (miRNA) expression in combination with ASCs therapy (unpublished data). MiRNAs are a class of small non-coding RNAs that function as repressors of gene expression at the level of post-transcriptional regulation, and the roles of miRNA on bone formation and homeostasis have been investigated (for review, see [117]). We construct multiple baculoviruses harboring miRNAs putatively associated with osteogenesis (e.g. miR-26a, miR-29b, miR-148b and miR-196a) for transduction of human ASCs, and unveil that baculovirus-mediated miR-148b and miR-196a overexpression more effectively up-regulate the osteogenic marker gene expression in human ASCs cultured in osteogenic medium. Co-transduction of human ASCs with baculovirus vectors expressing miR-148b and BMP-2 not only extends the BMP-2 expression but also augments the ASCs osteogenesis. Implantation of the baculovirus-engineered ASCs that express miR-148b and BMP-2 into the critical-size calvarial defects in nude mice remarkably ameliorates the bone regeneration (unpublished data).

Additionally, Li et al. [91] have employed miRNA to regulate the angiogenesis-osteogenesis coupling by transfecting human BMSCs with a miR-26a agomer (chemically modified single stranded RNA). The miR-26a over-expression in BMSCs significantly promotes both angiogenesis and osteogenesis in vitro, as judged by the significant upregulation of genes associated with osteogenesis (e.g. *runx2*, *col I*, *bmp2*) and angiogenesis (e.g. *vegf* and *ang1*). Conversely, transfection of BMSCs with the miR-26a inhibitor suppresses the in vitro osteogenesis and angiogenesis gene expression. Transplantation of the miR-26a-transfected BMSCs/hydrogel into a 5-mm calvarial bone defect in nude mice leads to enhanced in vivo miR-26a expression and results in complete repair of critical-size calvarial bone defect, which is accompanied by increased vascularization [91]. These data confirm that miR-26a promotes blood vessel and bone formation during calvarial defect repair [91]. Interestingly, miR-26a does not exert similarly potent osteoinductive effects in human ASCs in our study (unpublished data), probably due to the differences in the cell source.

In a more recent study, miR-31, a pleiotropically acting miRNA that inhibits cancer metastasis and targets special AT-rich sequence-binding protein 2 (SATB2) in fibroblasts, is shown to regulate ASCs osteogenesis and bone formation [90]. Rat ASCs are transduced with the lentiviral vector expressing miR-31 or miR-31 anti-sense RNA, seeded to β-tricalcium phosphate scaffold and implanted into critical-size calvarial defects in rats. The lentiviral vector expressing miR-31 anti-sense RNA significantly enhances osteogenic mRNA and protein expression, and a RunX2, SATB2 and miR-31 regulatory loop triggered by BMP-2 plays an important role in ASCs' osteogenic differentiation and bone regeneration. Furthermore, ASCs with miR-31 knock-down remarkably improve the repair of the calvarial defects, as evidenced by increased bone volume, elevated bone mineral density and decreased scaffold residue in vivo [90].

Besides these cells, the potential of induced pluripotent stem cells (iPSCs) in bone engineering is also assessed. It is first shown that transduction of iPSCs with an adenovirus expressing RunX2 enhances the osteogenesis in vitro [118].

Embryoid bodies derived from murine iPSCs cultured in differentiation medium for 12 weeks can also differentiate into osteoblasts [119]. These iPSCs-derived osteoblasts are seeded in a gelfoam matrix and implanted subcutaneously into syngeneic ICR mice, which gives rise to mineralized bone tissue with vascular supply in vivo [119]. To enhance the osteogenic differentiation, iPSCs are transduced with a retrovirus expressing a potent transcription factor, nuclear matrix protein SATB2 [72], which facilitates the osteogenic differentiation of iPSCs in vitro. Transplantation of the SATB2-overexpressing iPSCs together with silk scaffolds into critical-size calvarial bone defects in nude mice leads to enhanced new bone formation, thus demonstrating the feasibility of genetically modified iPSCs in bone tissue engineering.

## References


1. Lu C-H, Chang Y-H, Lin S-Y, Li K-C, Hu Y-C (2013) Recent progresses in gene delivery-based bone tissue engineering. Biotechnol Adv 31:1695–1706
2. Baltzer AWA, Lattermann C, Whalen JD, Wooley P, Weiss K, Grimm M et al (2000) Genetic enhancement of fracture repair: healing of an experimental segmental defect by adenoviral transfer of the BMP-2 gene. Gene Ther 7:734–739
3. Betz OB, Betz VM, Nazarian A, Pilapil CG, Vrahas MS, Bouxsein ML et al (2006) Direct percutaneous gene delivery to enhance healing of segmental bone defects. J Bone Joint Surg Am 88A:355–365
4. Betz VM, Betz OB, Glatt V, Gerstenfeld LC, Einhorn TA, Bouxsein ML et al (2007) Healing of segmental bone defects by direct percutaneous gene delivery: effect of vector dose. Hum Gene Ther 18:907–915
5. Betz OB, Betz VM, Nazarian A, Egermann M, Gerstenfeld LC, Einhorn TA et al (2007) Delayed administration of adenoviral BMP-2 vector improves the formation of bone in osseous defects. Gene Ther 14:1039–1044
6. Ishihara A, Shields KM, Litsky AS, Mattoon JS, Weisbrode SE, Bartlett JS et al (2008) Osteogenic gene regulation and relative acceleration of healing by adenoviral-mediated transfer of human BMP-2 or -6 in equine osteotomy and ostectomy models. J Orthop Res 26:764–771
7. Egermann M, Lill CA, Griesbeck K, Evans CH, Robbins PD, Schneider E et al (2006) Effect of BMP-2 gene transfer on bone healing in sheep. Gene Ther 13:1290–1299
8. Bertone AL, Pittman DD, Bouxsein ML, Li J, Clancy B, Seeherman HJ (2004) Adenoviral-mediated transfer of human BMP-6 gene accelerates healing in a rabbit ulnar osteotomy model. J Orthop Res 22:1261–1270
9. Alden TD, Beres EJ, Laurent JS, Engh JA, Das S, London SD et al (2000) The use of bone morphogenetic protein gene therapy in craniofacial bone repair. J Craniofac Surg 11:24–30
10. Jane JA, Dunford BA, Kron A, Pittman DD, Sasaki T, Li JZ et al (2002) Ectopic osteogenesis using adenoviral bone morphogenetic protein (BMP)-4 and BMP-6 gene transfer. Mol Ther 6:464–470
11. Liu YG, Zhou Y, Hu X, Fu JJ, Pan Y, Chu TW (2011) Effect of vascular endothelial growth factor 121 adenovirus transduction in rabbit model of femur head necrosis. J Trauma 70:1519–1523
12. Rundle CH, Miyakoshi N, Kasukawa Y, Chen ST, Sheng MHC, Wergedal JE et al (2003) In vivo bone formation in fracture repair induced by direct retroviral-based gene therapy with bone morphogenetic protein-4. Bone 32:591–601

13. Rundle CH, Strong DD, Chen ST, Linkhart TA, Sheng MHC, Wergedal JE et al (2008) Retroviral-based gene therapy with cyclooxygenase-2 promotes the union of bony callus tissues and accelerates fracture heating in the rat. J Gene Med 10:229–241
14. Lin L, Shen Q, Leng H, Duan X, Fu X, Yu C (2011) Synergistic inhibition of endochondral bone formation by silencing HIF-1α and Runx2 in trauma-induced heterotopic ossification. Mol Ther 19:1426–1432
15. Gafni Y, Pelled G, Zilberman Y, Turgeman G, Apparailly F, Yotvat H et al (2004) Gene therapy platform for bone regeneration using an exogenously regulated, AAV-2-based gene expression system. Mol Ther 9:587–595
16. Kimelman-Bleich N, Pelled G, Zilberman Y, Kallai I, Mizrahi O, Tawackoli W et al (2011) Targeted gene-and-host progenitor cell therapy for nonunion bone fracture repair. Mol Ther 19:53–59
17. Musgrave DS, Bosch P, Ghivizzani S, Robbins PD, Evans CH, Huard J (1999) Adenovirus-mediated direct gene therapy with bone morphogenetic protein-2 produces bone. Bone 24:541–547
18. Zou D, Zhang Z, Ye D, Tang A, Deng L, Han W et al (2011) Repair of critical-sized rat calvarial defects using genetically engineered BMSCs overexpressing HIF-1α. Stem Cells 29:1380–1390
19. Bright C, Park YS, Sieber AN, Kostuik JP, Leong KW (2006) In vivo evaluation of plasmid DNA encoding OP-1 protein for spine fusion. Spine (Phila Pa 1976) 31:2163–2172
20. Pelled G, Ben-Arav A, Hock C, Reynolds DG, Yazici C, Zilberman Y et al (2010) Direct gene therapy for bone regeneration: gene delivery, animal models, and outcome measures. Tissue Eng Part B Rev 16:13–20
21. Evans CH (2010) Gene therapy for bone healing. Expert Rev Mol Med 12:e18
22. Fang J, Zhu YY, Smiley E, Bonadio J, Rouleau JP, Goldstein SA et al (1996) Stimulation of new bone formation by direct transfer of osteogenic plasmid genes. Proc Natl Acad Sci U S A 93:5753–5758
23. Bonadio J, Smiley E, Patil P, Goldstein S (1999) Localized, direct plasmid gene delivery in vivo: prolonged therapy results in reproducible tissue regeneration. Nat Med 5:753–759
24. Backstrom KC, Bertone AL, Wisner ER, Weisbrode SE (2004) Response of induced bone defects in horses to collagen matrix containing the human parathyroid hormone gene. Am J Vet Res 65:1223–1232
25. Geiger F, Bertram H, Berger I, Lorenz H, Wall O, Eckhardt C et al (2005) Vascular endothelial growth factor gene-activated matrix (VEGF165-GAM) enhances osteogenesis and angiogenesis in large segmental bone defects. J Bone Miner Res 20:2028–2035
26. Endo M, Kuroda S, Kondo H, Maruoka Y, Ohya K, Kasugai S (2006) Bone regeneration by modified gene-activated matrix: effectiveness in segmental tibial defects in rats. Tissue Eng 12:489–497
27. Keeney M, van den Beucken JJ, van der Kraan PM, Jansen JA, Pandit A (2010) The ability of a collagen/calcium phosphate scaffold to act as its own vector for gene delivery and to promote bone formation via transfection with VEGF(165). Biomaterials 31:2893–2902
28. Itaka K, Ohba S, Miyata K, Kawaguchi H, Nakamura K, Takato T et al (2007) Bone regeneration by regulated in vivo gene transfer using biocompatible polyplex nanomicelles. Mol Ther 15:1655–1662
29. Huang YC, Simmons C, Kaigler D, Rice KG, Mooney DJ (2005) Bone regeneration in a rat cranial defect with delivery of PEI-condensed plasmid DNA encoding for bone morphogenetic protein-4 (BMP-4). Gene Ther 12:418–426
30. Chew SA, Kretlow JD, Spicer PP, Edwards AW, Baggett LS, Tabata Y et al (2011) Delivery of plasmid DNA encoding bone morphogenetic protein-2 with a biodegradable branched polycationic polymer in a critical-size rat cranial defect model. Tissue Eng Part A 17:751–763
31. Zhang W, Tsurushima H, Oyane A, Yazaki Y, Sogo Y, Ito A et al (2011) BMP-2 gene-fibronectin-apatite composite layer enhances bone formation. J Biomed Sci 18:62

32. Wang X, Oyane A, Tsurushima H, Sogo Y, Li X, Ito A (2011) BMP-2 and ALP gene expression induced by a BMP-2 gene-fibronectin-apatite composite layer. Biomed Mater 6:045004
33. Zhang Y, Fan W, Nothdurft L, Wu C, Zhou Y, Crawford R et al (2011) In vitro and in vivo evaluation of adenovirus combined silk fibroin scaffolds for bone morphogenetic protein-7 gene delivery. Tissue Eng Part C Methods 17:789–797
34. Zhang Y, Cheng N, Miron R, Shi B, Cheng X (2012) Delivery of PDGF-B and BMP-7 by mesoporous bioglass/silk fibrin scaffolds for the repair of osteoporotic defects. Biomaterials 33:6698–6708
35. Zhang Y, Shi B, Li C, Wang Y, Chen Y, Zhang W et al (2009) The synergetic bone-forming effects of combinations of growth factors expressed by adenovirus vectors on chitosan/collagen scaffolds. J Control Release 136:172–178
36. Luo T, Zhang W, Shi B, Cheng X, Zhang Y (2012) Enhanced bone regeneration around dental implant with bone morphogenetic protein 2 gene and vascular endothelial growth factor protein delivery. Clin Oral Implants Res 23:467–473
37. Lu SS, Zhang X, Soo C, Hsu T, Napoli A, Aghaloo T et al (2007) The osteoinductive properties of Nell-1 in a rat spinal fusion model. Spine J 7:50–60
38. Evans CH, Liu FJ, Glatt V, Hoyland JA, Kirker-Head C, Walsh A et al (2009) Use of genetically modified muscle and fat grafts to repair defects in bone and cartilage. Eur Cell Mater 18:96–111
39. Liu F, Porter RM, Wells J, Glatt V, Pilapil C, Evans CH (2012) Evaluation of BMP-2 gene-activated muscle grafts for cranial defect repair. J Orthop Res 30:1095–1102
40. Ito H, Koefoed M, Tiyapatanaputi P, Gromov K, Goater JJ, Carmouche J et al (2005) Remodeling of cortical bone allografts mediated by adherent rAAV-RANKL and VEGF gene therapy. Nat Med 11:291–297
41. Koefoed M, Ito H, Gromov K, Reynolds DG, Awad HA, Rubery PT et al (2005) Biological effects of rAAV-caAlk2 coating on structural allograft healing. Mol Ther 12:212–218
42. Yazici C, Takahata M, Reynolds DG, Xie C, Samulski RJ, Samulski J et al (2011) Self-complementary AAV2.5-BMP2-coated femoral allografts mediated superior bone healing versus live autografts in mice with equivalent biomechanics to unfractured femur. Mol Ther 19:1416–1425
43. Ben Arav A, Pelled G, Zilberman Y, Kimelman-Bleich N, Gazit Z, Schwarz EM et al (2012) Adeno-associated virus-coated allografts: a novel approach for cranioplasty. J Tissue Eng Regen Med 6:e43–e50
44. Chen F, Zhang X, Sun S, Zara JN, Zou X, Chiu R et al (2011) NELL-1, an osteoinductive factor, is a direct transcriptional target of Osterix. PLoS One 6:e24638
45. Chang PC, Cirelli JA, Jin Q, Seol YJ, Sugai JV, D'Silva NJ et al (2009) Adenovirus encoding human platelet-derived growth factor-B delivered to alveolar bone defects exhibits safety and biodistribution profiles favorable for clinical use. Hum Gene Ther 20:486–496
46. Evans CH (2012) Gene delivery to bone. Adv Drug Deliv Rev 64:1331–1340
47. Phinney DG, Prockop DJ (2007) Concise review: mesenchymal stem/multipotent stromal cells: the state of transdifferentiation and modes of tissue repair – current views. Stem Cells 25:2896–2902
48. Granero-Molto F, Weis JA, Miga MI, Landis B, Myers TJ, O'Rear L et al (2009) Regenerative effects of transplanted mesenchymal stem cells in fracture healing. Stem Cells 27:1887–1898
49. Kumar S, Wan C, Ramaswamy G, Clemens TL, Ponnazhagan S (2010) Mesenchymal stem cells expressing osteogenic and angiogenic factors synergistically enhance bone formation in a mouse model of segmental bone defect. Mol Ther 18:1026–1034
50. Gao J, Dennis JE, Muzic RF, Lundberg M, Caplan AI (2001) The dynamic in vivo distribution of bone marrow-derived mesenchymal stem cells after infusion. Cells Tissues Organs 169:12–20
51. Prockop DJ (2009) Repair of tissues by adult stem/progenitor cells (MSCs): controversies, myths, and changing paradigms. Mol Ther 17:939–946

52. Lien CY, Chih-Yuan Ho K, Lee OK, Blunn GW, Su Y (2009) Restoration of bone mass and strength in glucocorticoid-treated mice by systemic transplantation of CXCR4 and cbfa-1 co-expressing mesenchymal stem cells. J Bone Miner Res 24:837–848
53. Cho SW, Sun HJ, Yang J-Y, Jung JY, An JH, Cho HY et al (2009) Transplantation of mesenchymal stem cells overexpressing RANK-Fc or CXCR4 prevents bone loss in ovariectomized mice. Mol Ther 17:1979–1987
54. Guan M, Yao W, Liu R, Lam KS, Nolta J, Jia J et al (2012) Directing mesenchymal stem cells to bone to augment bone formation and increase bone mass. Nat Med 18:456–462
55. Hall SL, Lau K-HW, Chen S-T, Wergedal JE, Srivastava A, Klamut H et al (2007) Sca-1+ hematopoietic cell-based gene therapy with a modified FGF-2 increased endosteal/trabecular bone formation in mice. Mol Ther 15:1881–1889
56. Phillips JE, Gersbach CA, Garcia AJ (2007) Virus-based gene therapy strategies for bone regeneration. Biomaterials 28:211–229
57. Evans CH, Ghivizzani SC, Robbins PD (2009) Orthopedic gene therapy in 2008. Mol Ther 17:231–244
58. Kimelman N, Pelled G, Helm GA, Huard J, Schwarz EM, Gazit D (2007) Review: gene- and stem cell-based therapeutics for bone regeneration and repair. Tissue Eng 13:1135–1150
59. Moutsatsos IK, Turgeman G, Zhou SH, Kurkalli BG, Pelled G, Tzur L et al (2001) Exogenously regulated stem cell-mediated gene therapy for bone regeneration. Mol Ther 3:449–461
60. Sheyn D, Ruthemann M, Mizrahi O, Kallai I, Zilberman Y, Tawackoli W et al (2010) Genetically modified mesenchymal stem cells induce mechanically stable posterior spine fusion. Tissue Eng Part A 16:3679–3686
61. Sheyn D, Pelled G, Zilberman Y, Talasazan F, Frank JM, Gazit D et al (2008) Nonvirally engineered porcine adipose tissue-derived stem cells: use in posterior spinal fusion. Stem Cells 26:1056–1064
62. Sheyn D, Kallai I, Tawackoli W, Yakubovich DC, Oh A, Su SS et al (2011) Gene-modified adult stem cells regenerate vertebral bone defect in a rat model. Mol Pharm 8:1592–1601
63. Edwards PC, Ruggiero S, Fantasia J, Burakoff R, Moorji SM, Paric E et al (2005) Sonic hedgehog gene-enhanced tissue engineering for bone regeneration. Gene Ther 12:75–86
64. Byers BA, Guldberg RE, Hutmacher DW, Garcia AJ (2006) Effects of Runx2 genetic engineering and in vitro maturation of tissue-engineered constructs on the repair of critical size bone defects. J Biomed Mater Res A 76:646–655
65. Gysin R, Wergedal JE, Sheng MH, Kasukawa Y, Miyakoshi N, Chen ST et al (2002) Ex vivo gene therapy with stromal cells transduced with a retroviral vector containing the BMP4 gene completely heals critical size calvarial defect in rats. Gene Ther 9:991–999
66. Lee CW, Martinek V, Usas A, Musgrave D, Pickvance EA, Robbins P et al (2002) Muscle-based gene therapy and tissue engineering for treatment of growth plate injuries. J Pediatr Orthop 22:565–572
67. Wright VJ, Peng HR, Usas A, Young B, Gearhart B, Cummins J et al (2002) BMP4-expressing muscle-derived stem cells differentiate into osteogenic lineage and improve bone healing in immunocompetent mice. Mol Ther 6:169–178
68. Peng H, Usas A, Gearhart B, Olshanski A, Shen HC, Huard J (2004) Converse relationship between in vitro osteogenic differentiation and in vivo bone healing elicited by different populations of muscle-derived cells genetically engineered to express BMP4. J Bone Miner Res 19:630–641
69. Peng H, Wright V, Usas A, Gearhart B, Shen HC, Cummins J et al (2002) Synergistic enhancement of bone formation and healing by stem cell-expressed VEGF and bone morphogenetic protein-4. J Clin Invest 110:751–759
70. Peng H, Usas A, Hannallah D, Olshanski A, Cooper GM, Huard J (2005) Noggin improves bone healing elicited by muscle stem cells expressing inducible BMP4. Mol Ther 12:239–246
71. Peng H, Usas A, Olshanski A, Ho AM, Gearhart B, Cooper GM et al (2005) VEGF improves, whereas sFlt1 inhibits, BMP-2 induced bone formation and bone healing through modulation of angiogenesis. J Bone Miner Res 20:2017–2027

72. Ye J-H, Xu Y-J, Gao J, Yan S-G, Zhao J, Tu Q et al (2011) Critical-size calvarial bone defects healing in a mouse model with silk scaffolds and SATB2-modified iPSCs. Biomaterials 32:5065–5076
73. Turgeman G, Pittman DD, Muller R, Kurkalli BG, Zhou SH, Pelled G et al (2001) Engineered human mesenchymal stem cells: a novel platform for skeletal cell mediated gene therapy. J Gene Med 3:240–251
74. Blum JS, Barry MA, Mikos AG, Jansen JA (2003) In vivo evaluation of gene therapy vectors in ex vivo-derived marrow stromal cells for bone regeneration in a rat critical-size calvarial defect model. Hum Gene Ther 14:1689–1701
75. Xu XL, Tang T, Dai K, Zhu Z, Guo XE, Yu C et al (2005) Immune response and effect of adenovirus-mediated human BMP-2 gene transfer on the repair of segmental tibial bone defects in goats. Acta Orthop 76:637–646
76. Peterson B, Zhang J, Iglesias R, Kabo M, Hedrick M, Benhaim P et al (2005) Healing of critically sized femoral defects, using genetically modified mesenchymal stem cells from human adipose tissue. Tissue Eng 11:120–129
77. Park J, Ries J, Gelse K, Kloss F, von der Mark K, Wiltfang J et al (2003) Bone regeneration in critical size defects by cell-mediated BMP-2 gene transfer: a comparison of adenoviral vectors and liposomes. Gene Ther 10:1089–1098
78. Lieberman JR, Daluiski A, Stevenson S, Wu L, McAllister P, Lee YP et al (1999) The effect of regional gene therapy with bone morphogenetic protein-2-producing bone-marrow cells on the repair of segmental femoral defects in rats. J Bone Joint Surg Am 81A:905–917
79. Zhang XP, Xie C, Lin ASP, Ito H, Awad H, Lieberman JR et al (2005) Periosteal progenitor cell fate in segmental cortical bone graft transplantations: implications for functional tissue engineering. J Bone Miner Res 20:2124–2137
80. Steinhardt Y, Aslan H, Regev E, Zilberman Y, Kallai I, Gazit D et al (2008) Maxillofacial-derived stem cells regenerate critical mandibular bone defect. Tissue Eng Part A 14:1763–1773
81. Zhao Z, Wang Z, Ge C, Krebsbach P, Franceschi RT (2007) Healing cranial defects with AdRunx2-transduced marrow stromal cells. J Dent Res 86:1207–1211
82. Lattanzi W, Parrilla C, Fetoni A, Logroscino G, Straface G, Pecorini G et al (2008) Ex vivo-transduced autologous skin fibroblasts expressing human Lim mineralization protein-3 efficiently form new bone in animal models. Gene Ther 15:1330–1343
83. He X, Dziak R, Yuan X, Mao K, Genco R, Swihart M et al (2013) BMP2 genetically engineered MSCs and EPCs promote vascularized bone regeneration in rat critical-sized calvarial bone defects. PLoS One 8:e60473
84. Feeley BT, Conduah AH, Sugiyama O, Krenek L, Chen ISY, Lieberman JR (2006) In vivo molecular imaging of adenoviral versus lentiviral gene therapy in two bone formation models. J Orthop Res 24:1709–1721
85. Virk MS, Conduah A, Park SH, Liu N, Sugiyama O, Cuomo A et al (2008) Influence of short-term adenoviral vector and prolonged lentiviral vector mediated bone morphogenetic protein-2 expression on the quality of bone repair in a rat femoral defect model. Bone 42:921–931
86. Hsu WK, Sugiyama O, Park SH, Conduah A, Feeley BT, Liu NQ et al (2007) Lentiviral-mediated BMP-2 gene transfer enhances healing of segmental femoral defects in rats. Bone 40:931–938
87. Virk MS, Sugiyama O, Park SH, Gambhir SS, Adams DJ, Drissi H et al (2011) "Same Day" ex-vivo regional gene therapy: a novel strategy to enhance bone repair. Mol Ther 19:960–968
88. Zou D, Zhang Z, He J, Zhu S, Wang S, Zhang W et al (2011) Repairing critical-sized calvarial defects with BMSCs modified by a constitutively active form of hypoxia-inducible factor-1alpha and a phosphate cement scaffold. Biomaterials 32:9707–9718
89. Zou D, Zhang Z, He J, Zhang K, Ye D, Han W et al (2012) Blood vessel formation in the tissue-engineered bone with the constitutively active form of HIF-1α mediated BMSCs. Biomaterials 33:2097–2108
90. Deng Y, Zhou H, Zou D, Xie Q, Bi X, Gu P et al (2013) The role of miR-31-modified adipose tissue-derived stem cells in repairing rat critical-sized calvarial defects. Biomaterials 34:6717–6728

91. Li Y, Fan L, Liu S, Liu W, Zhang H, Zhou T et al (2013) The promotion of bone regeneration through positive regulation of angiogenic-osteogenic coupling using microRNA-26a. Biomaterials 34:5048–5058
92. Chuang C-K, Lin K-J, Lin C-Y, Chang Y-H, Yen T-C, Hwang S-M et al (2010) Xenotransplantation of human mesenchymal stem cells into immunocompetent rats for calvarial bone repair. Tissue Eng Part A 16:479–488
93. Lin C-Y, Chang Y-H, Lin K-J, Yen T-Z, Tai C-L, Chen C-Y et al (2010) The healing of critical-sized femoral segmental bone defects in rabbits using baculovirus-engineered mesenchymal stem cells. Biomaterials 31:3222–3230
94. Lin C-Y, Lin K-J, Kao C-Y, Chen M-C, Yen T-Z, Lo W-H et al (2011) The role of adipose-derived stem cells engineered with the persistently expressing hybrid baculovirus in the healing of massive bone defects. Biomaterials 32:6505–6514
95. Lin C-Y, Lin K-J, Li K-C, Sung L-Y, Hsueh S, Lu C-H et al (2012) Immune responses during healing of massive segmental femoral bone defects mediated by hybrid baculovirus-engineered ASCs. Biomaterials 33:7422–7434
96. Lin C-Y, Chang Y-H, Kao C-Y, Lu C-H, Sung L-Y, Yen T-C et al (2012) Augmented healing of critical-size calvarial defects by baculovirus-engineered MSCs that persistently express growth factors. Biomaterials 33:3682–3692
97. Szpalski C, Barr J, Wetterau M, Saadeh PB, Warren SM (2010) Cranial bone defects: current and future strategies. Neurosurg Focus 29:E8
98. Patel ZS, Young S, Tabata Y, Jansen JA, Wong ME, Mikos AG (2008) Dual delivery of an angiogenic and an osteogenic growth factor for bone regeneration in a critical size defect model. Bone 43:931–940
99. Young S, Patel ZS, Kretlow JD, Murphy MB, Mountziaris PM, Baggett LS et al (2009) Dose effect of dual delivery of vascular endothelial growth factor and bone morphogenetic protein-2 on bone regeneration in a rat critical-size defect model. Tissue Eng Part A 15:2347–2362
100. Geuze RE, Theyse LF, Kempen DH, Hazewinkel HA, Kraak HY, Oner FC et al (2012) A differential effect of bone morphogenetic protein-2 and vascular endothelial growth factor release timing on osteogenesis at ectopic and orthotopic sites in a large-animal model. Tissue Eng Part A 18:2052–2062
101. Kempen DHR, Lu L, Heijink A, Hefferan TE, Creemers LB, Maran A et al (2009) Effect of local sequential VEGF and BMP-2 delivery on ectopic and orthotopic bone regeneration. Biomaterials 30:2816–2825
102. Li GH, Corsi-Payne K, Zheng B, Usas A, Peng HR, Huard J (2009) The dose of growth factors influences the synergistic effect of vascular endothelial growth factor on bone morphogenetic protein 4-induced ectopic bone formation. Tissue Eng Part A 15:2123–2133
103. Helmrich U, Di Maggio N, Guven S, Groppa E, Melly L, Largo RD et al (2013) Osteogenic graft vascularization and bone resorption by VEGF-expressing human mesenchymal progenitors. Biomaterials 34:5025–5035
104. Jabbarzadeh E, Starnes T, Khan YM, Jiang T, Wirtel AJ, Deng M et al (2008) Induction of angiogenesis in tissue-engineered scaffolds designed for bone repair: a combined gene therapy-cell transplantation approach. Proc Natl Acad Sci 105:11099–11104
105. Chen C-Y, Lin C-Y, Chen G-Y, Hu Y-C (2011) Baculovirus as a gene delivery vector: recent understandings of molecular alterations in transduced cells and latest applications. Biotechnol Adv 29:618–631
106. Lin C-Y, Lu C-H, Luo W-Y, Chang Y-H, Sung L-Y, Chiu H-Y et al (2010) Baculovirus as a gene delivery vector for cartilage and bone tissue engineering. Curr Gene Ther 10:242–254
107. Ho Y-C, Chung Y-C, Hwang S-M, Wang K-C, Hu Y-C (2005) Transgene expression and differentiation of baculovirus-transduced human mesenchymal stem cells. J Gene Med 7:860–868
108. Allay JA, Dennis JE, Haynesworth SE, Majumdar MK, Clapp DW, Shultz LD et al (1997) LacZ and interleukin-3 expression in vivo after retroviral transduction of marrow-derived human osteogenic mesenchymal progenitors. Hum Gene Ther 8:1417–1427

109. Tsuda H, Wada T, Ito Y, Uchida H, Dehari H, Nakamura K et al (2003) Efficient BMP2 gene transfer and bone formation of mesenchymal stem cells by a fiber-mutant adenoviral vector. Mol Ther 7:354–365
110. Lo W-H, Hwang S-M, Chuang C-K, Chen C-Y, Hu Y-C (2009) Development of a hybrid baculoviral vector for sustained transgene expression. Mol Ther 17:658–666
111. Chuang C-K, Sung L-Y, Hwang S-M, Lo W-H, Chen H-C, Hu Y-C (2007) Baculovirus as a new gene delivery vector for stem cells engineering and bone tissue engineering. Gene Ther 14:1417–1424
112. Prigozhina TB, Khitrin S, Elkin G, Eizik O, Morecki S, Slavin S (2008) Mesenchymal stromal cells lose their immunosuppressive potential after allotransplantation. Exp Hematol 36:1370–1376
113. Niemeyer P, Kornacker M, Mehlhorn A, Seckinger A, Vohrer J, Schmal H et al (2007) Comparison of immunological properties of bone marrow stromal cells and adipose tissue-derived stem cells before and after osteogenic differentiation in vitro. Tissue Eng 13:111–121
114. Niemeyer P, Fechner K, Milz S, Richter W, Suedkamp NP, Mehlhorn AT et al (2010) Comparison of mesenchymal stem cells from bone marrow and adipose tissue for bone regeneration in a critical size defect of the sheep tibia and the influence of platelet-rich plasma. Biomaterials 31:3572–3579
115. Scotti C, Tonnarelli B, Papadimitropoulos A, Scherberich A, Schaeren S, Schauerte A et al (2010) Recapitulation of endochondral bone formation using human adult mesenchymal stem cells as a paradigm for developmental engineering. Proc Natl Acad Sci U S A 107: 7251–7256
116. Lin C-Y, Chang Y-H, Li K-C, Lu C-H, Sung L-Y, Yeh C-L et al (2013) The use of ASCs engineered to express BMP2 or TGF-β3 within scaffold constructs to promote calvarial bone repair. Biomaterials 34:9401–9412
117. Lian JB, Stein GS, van Wijnen AJ, Stein JL, Hassan MQ, Gaur T et al (2012) MicroRNA control of bone formation and homeostasis. Nat Rev Endocrinol 8:212–227
118. Tashiro K, Inamura M, Kawabata K, Sakurai F, Yamanishi K, Hayakawa T et al (2009) Efficient adipocyte and osteoblast differentiation from mouse induced pluripotent stem cells by adenoviral transduction. Stem Cells 27:1802–1811
119. Bilousova G, Jun DH, King KB, De Langhe S, Chick WS, Torchia EC et al (2011) Osteoblasts derived from induced pluripotent stem cells form calcified structures in scaffolds both in vitro and in vivo. Stem Cells 29:206–216

# Chapter 4
# Gene Therapy for Cartilage Tissue Engineering

**Abstract** Gene therapy has converged with cartilage engineering in recent years, by which an increasing number of therapeutic genes have been explored to stimulate cartilage repair. These genes can be administered to cells via in vivo or ex vivo approaches using either viral or nonviral vectors. This chapter reviews various growth factors and delivery approaches under investigation.

## 4.1 Gene Products for Promoting Chondrogenesis

A panel of growth factors capable of inducing chondrogenesis has been identified. These growth factors include insulin-like growth factor 1 (IGF-1), bone morphogenetic proteins (BMPs) such as BMP-2 [1], BMP-4 [2, 3] and BMP-6 [4, 5]. Moreover, transforming growth factors (TGF) including TGF-β1 [6, 7], TGF-β2 [8], TGF-β3 [9, 10] as well as growth and differentiation factor-5 (GDF5) [11] have been used to induce the chondrogenesis of stem cells.

Among these chondroinductive growth factors, TGF-β3 has gained popularity in recent years. In one study, BMSCs are seeded into a hybrid scaffold containing TGF-β3 and the constructs are implanted in rabbits for the repair of chondral defects [12]. After 8 weeks, differentiated BMSCs are located in lacunae within the matrix and exhibit typical chondrocyte morphology. Importantly, TGF-β3 induction of mesenchymal stem cells (MSCs) [13] and adipose-derived stem cells (ASCs) [14] is reported to more effective than TGF-β1 and TGF-β2, and controlled release of TGF-β3 may inhibit the osteogenesis of human MSCs [15]. The optimal amount of TGF-β3 remains to be established but very high doses of TGF-β3 (e.g. >900 ng/ml) are associated with synovitis, pannus formation, cartilage erosion and joint effusion [16].

Y.-C. Hu, *Gene Therapy for Cartilage and Bone Tissue Engineering*, SpringerBriefs in Bioengineering, DOI 10.1007/978-3-642-53923-7_4,

In addition to the use of single growth factor, cocktails of BMP and TGF-β have been investigated for their synergistic chondroinductive effects. The growth factor cocktails that can enhance chondrogenesis of ASCs/BMSCs include BMP-2+TGF-β1 [17], BMP-2+TGF-β3 [18], BMP-7+TGF-β2 [19], BMP-7+FGF-2 [20] and BMP-6+TGF-β3 [10]. Among these growth factor recipes, TGF-β3 in conjunction with BMP-6 seems to impose the most potent chondroinductive effect for ASCs [21] because BMP-6 can synergize the chondroinductive effect of TGF-β3 by inducing the expression of TGF-β receptor I which is usually not expressed by ASCs [10]. However, chondrogenesis of ASCs induced with TGF-β3+BMP-6 is also associated with hypertrophy in vitro and calcification in vivo [10].

In addition to anabolic factors, other molecules capable of suppressing cartilage breakdown have been exploited. interleukin 1 (IL-1) is a proinflammatory cytokine contributing to the pathology of osteoarthritis and rheumatoid arthritis, while the IL-1 receptor agonist protein (IL-1Ra) may reverse cartilage loss in osteoarthritis [22]. Conversely, IL-10 has both immunostimulatory and immunosuppressive properties and a homologue of IL-10 encoded by Epstein-Barr virus (EBV), known as viral IL-10 (vIL-10), is also able to suppress the immune response. Nell-1 (NEL-like molecule-1) is a secreted molecule and is expressed preferentially in cells of neural crest origin residing within the craniofacial complex and central nervous system. It is a growth factor believed to specifically target cells committed to the osteochondral lineage [23]. Nell-1 can also promote chondrocytes proliferation and deposition of cartilage-specific extracellular matrix (ECM) in vitro [24], suggesting potential therapeutic benefits of Nell-1 in the stem cell-based repair of osteochondral defect. In addition, sex-determining region Y box gene 9 (SOX-9) is a transcription factor that can activate chondrocyte-specific enhancer elements in the *col2a1*, *col9a1*, *col11a2*, and *aggrecan* genes. Therefore, SOX-9 is a "master regulator" of the chondrocyte phenotype and SOX-9 is expressed in all chondroprogenitor cells, predominantly in mesenchymal condensations and cartilage [25]. SOX-9 can effectively induce chondrogenesis of BMSCs both in monolayer and on the polymeric scaffold [26]. Two other members of the SOX family, SOX-5 and SOX-6, are also required for chondrogenesis. In vitro and in vivo studies show that SOX-5 and SOX-6 cooperate with SOX-9 to activate the *col2a1* enhancer in chondrogenic cells. Deletion of SOX-5 and SOX-6 in mice causes a severe, generalized chondrodysplasia [27].

## 4.2 In Vivo Gene Delivery-Based Cartilage Engineering

Synovium lines the internal surfaces of joint space and has a relatively large surface area. Because gene vectors that are injected to the joint space may directly transduce the cells in the synovial lining, direct intra-articular injection of a vector into the joint space is conceptually easy. In fact, direct intra-articular injection of retrovirus expressing a reporter gene results in transduction of synoviocytes and sustained reporter gene expression [28]. Also, intra-articular injection of AAV is capable of transducing articular chondrocytes in vivo [29]. Therefore, a number of studies have attempted to inject the gene vectors expressing various transgenes (Table 4.1).

**Table 4.1** In vivo gene therapy for cartilage engineering via direct injection

| Vector | Transgene | Model | References |
|---|---|---|---|
| Adenovirus | vIL-10 | Rabbit knee joints with induced arthritis | [30] |
| HSV | IL-1Ra | Rabbit knee joints with induced arthritis | [31] |
| AAV | IL-1Ra | Rat knee joints with induced arthritis | [32] |
| Lentivirus | IL-1Ra | Rat knee joints with induced arthritis | [33] |
| Adenovirus | IL-1Ra | Horse knee joints with experimental osteoarthritis | [34] |
| AAV | TNFR:Fc | Human patients with rheumatoid arthritis | [35, 36] |
| Adenovirus | IGF-1 | Rabbit knee knees joints (normal and arthritic) | [37] |
| Adenovirus | TGF-β1 | Rabbit knee joints with arthritis | [38] |
| AAV | IL-4 | Mouse knees joints with induced arthritis | [39] |
| AAV | FGF-2 | Rabbit knee joints | [40] |
| AAV | SOX-9 | Human arthritis cartilage | [41] |
| AAV | SOX-9 | Rabbit knee joints | [42] |

### *4.2.1 Strategies to Suppress Cartilage Degeneration*

Early studies have initially attempted to inhibit cartilage degeneration for the treatment of arthritis. It is shown that intra-articular injection of adenovirus expressing vIL-10 into the knee joints of rabbits with antigen-induced arthritis significantly reduces leukocytosis, degrees of synovitis, cartilage matrix degradation and levels of endogenous rabbit TNF-α, while maintaining high levels of cartilage matrix synthesis [30]. Intra-articular injection of herpes simplex virus (HSV) vector expressing IL-1R antagonist (IL-1Ra) into the knee joints of rabbits with experimental arthritis suppresses leukocytosis and synovitis significantly, demonstrating the feasibility of in vivo inflammation repression by IL-1Ra expression [31]. Furthermore, direct injection of an AAV encoding IL-1Ra into the rat joints with liposaccharide-induced arthritis can suppress primary and recurrent arthritis [32]. Delivery of an adenovirus expressing the equine homolog of IL-1Ra into the joints of horse with experimental osteoarthritis (OA) also results in significant clinical improvement in both pain and disease activity, and preservation of articular cartilage [34]. IL-1Ra is also delivered by lentivirus into the knee joins of rats [33], which strongly prevents swelling in all arthritic knees. Cellular infiltration, cartilage erosion, and invasiveness of inflamed synovium are effectively prevented in the knees treated with the lentiviral vector [43].

Since tumor necrosis factor-α (TNF-α) plays pivotal roles in rheumatoid arthritis (RA) pathogenesis [44] and soluble TNF receptor (TNFR) is an antagonist that counters TNF-α activities, biweekly administration of a TNFR-immunoglobulin Fc fusion protein (TNFR:Fc, etanercept) ameliorates RA joint symptoms [45]. Therefore, a single stranded AAV serotype 2 (AAV2) vector expressing the fusion protein (TNFR:Fc) is employed in phase I and II clinical trials aiming for the treatment of RA [35, 36]. The gene product is identical to etanercept (Enbrel®) used to treat patients with RA and blocks the actions of TNF [46]. The clinical study demonstrates symptomatic benefit in some patients [36], yet suffers from a major setback in 2007 as one patient died shortly after receiving a second intra-articular injection of AAV2 vector. However, the US Food and Drug Administration has concluded that the death is unrelated to the gene transfer [47].

### 4.2.2 Strategies to Promote Cartilage Formation

In contrast to inhibiting cartilage degeneration, strategies have been explored to stimulate cartilage regeneration. For instance, delivery of an adenovirus expressing IGF-1 to the normal and arthritic rabbit knees is able to stimulate an increase in proteoglycan synthesis in articular cartilage without adverse effects [37]. Direct injection of the AAV encoding chondroprotective IL-4 into the knee joints of mice with collagen-induced arthritis leads to detectable IL-4 expression in the joint, and protection of articular cartilage destruction [39]. Furthermore, direct injection of AAV encoding FGF-2 into the osteochondral defects in the patellar groove of rabbit knee joints improves the overall cartilage repair, filling, architecture and cell morphology of osteochondral defects [40]. Direct injection of AAV encoding SOX-9 into human osteoarthritis cartilage restores the production of proteoglycans and collagen II [41]. Furthermore, injection of AAV encoding SOX-9 into the osteochondral defects in rabbit knee joints is capable of improving cartilage repair processes with enhanced production of major matrix components [42]. The treatment also delays premature terminal differentiation and hypertrophy in the newly formed cartilage, possibly due to contrasting effects of SOX-9 on Runt-related transcription factor 2 (RunX2) and β-catenin osteogenic expression in this area. Strikingly, SOX-9 treatment improves the reconstitution of the subchondral bone in the defects, possibly due to an increase in RunX2 expression in this location [42].

Moreover, direct injection of adenovirus expressing TGF-β1 into the antigen-induced arthritic rabbit knee joints results in a dose-dependent TGF-β1 expression in the synovial fluid, induction of chondrogenesis within the synovial lining and suppression of inflammation [38]. However, it is also shown that injection of adenovirus expressing TGF-β1 is unable to stimulate repair of damaged cartilage and even triggers cartilage degradation, thus gene transfer of TGF-β1 to the synovium is suggested to be unsuitable for treating intra-articular pathologies [38] and a tightly coordinated regulation of TGF-β1 is needed to control chondrogenesis [48].

### 4.2.3 Problems in Direct Gene Transfer

Despite the promise and ease of intra-articular vector injection, it is suggested that synovium might not be ideal for the expression of pleiotropic protein by direct gene transfer because the gene products might stimulate undesired activities in non-target cells. Furthermore, direct vector injection may elicit inflammatory responses that interfere with the reparative process [49]. The dense matrix that surrounds chondrocytes imposes another barrier to direct, in vivo gene delivery to chondrocytes by nearly all vectors [50]. Although AAV can transduce cartilage explants in culture [51, 52] AAV transduces other cells in the synovium and does not specifically target chondrocytes following direct injection into the joint [53]. As a result, BMP-2 over-expression by direct gene transfer into the synovial lining of mouse knees causes the formation of ectopic cartilage throughout the joint capsule and growth of large osteophytes [54].

Another concern about the success of intra-articular injection is the inability to achieve long-term expression in animal models. It has been shown that pre-existing antibody against AAV within the synovial fluid and sera inhibits AAV transduction of chondrocytes [55]. Furthermore, in the rat knees the immune responses to heterologous transgene products and viral proteins diminish the transgene expression and result in synovial cell turnover [56].

## 4.3 Ex Vivo Gene Therapy

Another common approach involves ex vivo gene delivery into appropriate cell types, followed by encapsulation into a scaffold and implantation of the cells/scaffold constructs into the cartilage defect [57]. Alternatively, the transgene can immobilized onto the matrix, followed by the seeding of cells and implantation of the gene activated matrix (GAM) into the defect. The ex vivo approach minimizes the unwanted immune responses and is more popular than the in vivo approach. Selected examples of ex vivo gene therapy for cartilage engineering are summarized in Table 4.2.

### *4.3.1 Gene Transfer to Chondrocytes*

Since chondrocyte is the sole cell type within articular cartilage, it is tempting to improve the quality of repair tissue by ex vivo modification of chondrocytes when they are undergoing expansion in culture. Transduction of articular chondrocytes in vitro with an adenovirus expressing TGF-β1 enhances the synthesis of proteoglycan, collagen and noncollagenous proteins [69]. Transduction of articular chondrocytes with adenovirus expressing IGF-1 similarly increases matrix biosynthesis and maintains the chondrocyte phenotype [70]. Additionally, transduction of bovine chondrocytes with an adenovirus expressing BMP-7 and transplantation onto cartilage explants for in vitro culture enhances the chondrocyte-specific matrix synthesis and their capacity to form cartilage-like tissues [58]. Implantation of the chondrocytes expressing BMP-7 into the extensive articular cartilage defects in horses accelerates the appearance of hyaline-like repair tissue although only few implanted cells persist at 8 months post-implantation [58].

Furthermore, rabbit chondrocytes can be transduced with an AAV expressing basic fibroblast growth factor (bFGF). These transduced chondrocytes are embedded

**Table 4.2** Selected examples of ex vivo gene therapy for articular cartilage tissue engineering

| Vector | Cells | Transgene | Models | References |
|---|---|---|---|---|
| Plasmid | Articular chondrocytes | IGF-1 | Osteochondral defect in rabbit knees | [57] |
| Adenovirus | Articular chondrocytes | BMP-7 | Horse articular cartilage defect | [58] |
| AAV | Articular chondrocytes | bFGF | Rabbit articular cartilage defect | [59] |
| Baculovirus | Articular chondrocytes | BMP-2 | Rabbit osteochondral defect | [60] |
| Retrovirus | Synoviocyte | IL-10, IL-1Ra | Rabbit osteoarthritis model | [61] |
| Retrovirus | Synovial fibroblasts | IL-1Ra | Humans with arthritis | [62] |
| Plasmid | Periosteal stem cells | BMP-7, Shh | Rabbit osteochondral defect | [63] |
| Adenovirus | Perichondrium mesenchymal cells | BMP-2, IGF-1 | Rat partial-thickness cartilage defect | [64] |
| AAV | BMSCs | TGF-β1 | Rat osteochondral defect | [6] |
| Adenovirus | BMSCs | Nell-1 | Goat osteochondral defects at the mandibular condyle | [65] |
| Plasmid | BMSCs | SOX-9 | Ectopic, subcutaneous mouse model | [66] |
| Adenovirus | BMSCs | SOX-9 | Rabbit full-thickness cartilage defect | [26] |
| Adenovirus | ASCs | TGF-β2 | Mouse subcutaneous model | [8] |
| Plasmid | ASCs | SOX-5, -6, and -9 | Rat osteochondral defect and rat osteoarthritis model | [27] |
| Baculovirus | ASCs | TGF-β3/ BMP-6 | Rabbit full-thickness defect | [67] |
| Adenovirus | Stem cells from perichondrium/ periosteum, bone marrow and fat | BMP-2 | Rat partial-thickness defect | [68] |

in collagen gel and transplanted into a full-thickness defect in the articular cartilage at the patellar grooves in rabbits. The AAV vector results in prolonged expression for 8 weeks and improves the repair of rabbit articular cartilage [59].

In contrast to repair approaches based on formation and maturation of new tissue in situ, the generation of cartilaginous tissues can be achieved by implanting a preformed graft [49]. Implantation of a preformed construct may be preferred because the cell/scaffold constructs devoid of cartilage ECM lacks appropriate mechanical properties to tolerate the mechanical loading in vivo immediately after implantation, which could impair subsequent tissue integration [71]. In one of the most typical approaches to creating tissue engineered cartilage equivalents, cells are seeded into porous scaffolds and the cell/scaffold constructs are cultured in vitro in a bioreactor [72, 73]. Bioreactor systems provide mechanical stimuli and oxygen/nutrient transfer to promote chondrogenesis, extracellular matrix (ECM) synthesis and cartilage growth [74, 75]. In this regard, Madry et al. genetically modify primary chondrocytes via IGF-I cDNA transfection, seed the cells into polymer scaffolds and culture the constructs in the rotating wall bioreactor [76]. Four-week culture of the constructs leads to excellent glycosaminoglycans (GAGs) and collagen yield and the formation of cartilage-like tissues suitable for implantation. Implanting the cartilaginous constructs into osteochondral defects in rabbit knees repairs the articular cartilage defect and accelerates the formation of subchondral bone [57].

In addition, baculovirus transduces primary rat chondrocytes effectively and baculovirus DNA degrades with time gradually [77]. Although cell proliferation is slightly hindered after virus transduction due to the transgene expression, the cell growth rate restores after subculture and cease of transgene expression. Critically, the transduced chondrocytes retain the ability to deposit articular cartilage-specific collagen II and GAGs, showing that baculovirus transduction does not mitigate the normal differentiation state of rat chondrocytes [77]. Therefore, the baculovirus-transduced chondrocytes are seeded into porous polymeric scaffolds and cultivated in a rotating-shaft bioreactor (RSB) developed for two-phase cultivation of tissue engineered cartilage [78]. The baculovirus transduction affects neither cell adhesion to the porous scaffold nor cell survival in the RSB [79]. After 4-week culture in the RSB, the transduced chondrocytes remain highly differentiated and the cell/scaffold constructs grow into cartilage-like tissues that are indistinguishable from the untransduced controls, demonstrating that baculovirus transduction neither harms chondrocytes nor retards the formation of cartilage-like tissues in the RSB [79].

One obstacle to employing mature chondrocytes in cartilage tissue engineering is that the freshly isolated cells need to be serially passaged in order to expand the cell number, which however, results in progressive cell de-differentiation and loss of chondrocyte function. To tackle this problem, recombinant baculoviruses expressing TGF-β1 (Bac-CT), IGF-1 (Bac-CI) or BMP-2 (Bac-CB) are constructed [80] because TGF-β1, IGF-1 and BMP-2 can promote the synthesis and deposition of ECM by chondrocytes [81]. The baculovirus transduces rabbit articular chondrocytes that are subcultured to different passages: passage 1 (P1), passage 3 (P3) and passage 5 (P5). At a multiplicity of infection (MOI) of 50, baculovirus transduces rabbit chondrocytes at efficiencies ranging from 80 to 82 %, in a passage-independent manner, which are comparable or superior to those mediated by adenoviral [82], AAV [83] or retroviral vectors [84] at similar vector doses.

Transduction with Bac-CT, Bac-CI and Bac-CB leads to the expression of TGF-β1, IGF-1 and BMP-2 to therapeutic levels in rabbit chondrocytes, yet in a passage-dependent manner (Fig. 4.1). Albeit sufficient protein expression, the outcome hinges on the growth factor and cell passage [80]. The de-differentiated P5 chondrocytes fail to respond to the stimulation by either growth factor. The partially de-differentiated P3 cells also fail to maintain the chondrocyte phenotype. Nonetheless, baculovirus-mediated BMP-2 expression (Bac-CB transduction)

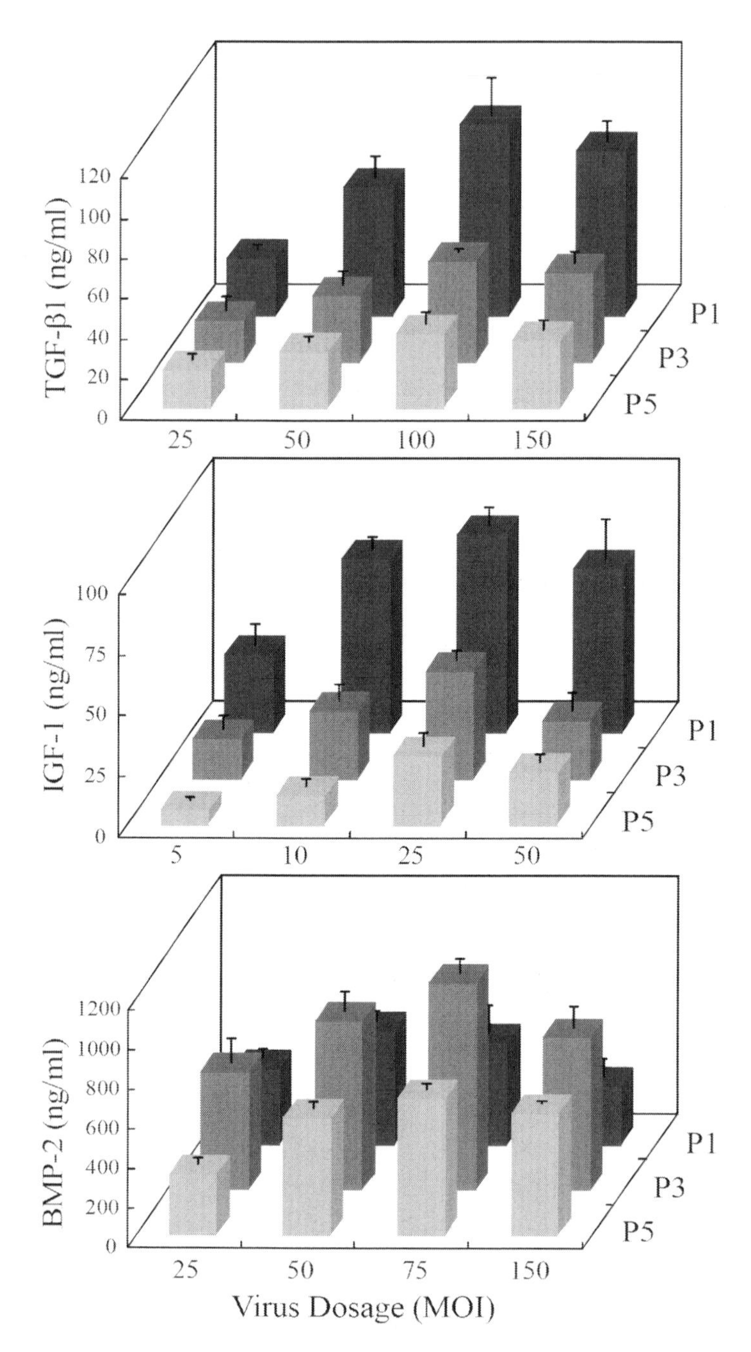


**Fig. 4.1 Effects of cell passage and virus dosage on growth factor expression**. The rabbit articular chondrocytes of different passages are transduced with Bac-CT (expressing TGF-β1), Bac-CI (expressing IGF-1) or Bac-CB (expressing BMP-2) at the indicated virus dosage. The concentrations of TGF-β1, IGF-1 and BMP-2 in the medium are assayed at 1 day post-transduction using ELISA kits.

remarkably reverses the de-differentiation and enhances the GAGs and collagen II production, as evidenced by cell morphology (Fig. 4.2a), histological staining (Fig. 4.2b) and gene expression analyses. Bac-CT modestly enhances the chondrogenesis but is insufficient to restore the differentiation of P3 cells, which is consistent with the finding that adenovirus-mediated expression of TGF-β1 alone can not rescue the collagen phenotype of passaged chondrocytes [69]. Intriguingly, IGF-1, a well-known chondroinductive protein, fails to stimulate the P3 cells likely due to the loss of IGF-1 receptor expression and hence the desensitization of IGF-1 stimulation [85]. This study highlights the importance of selecting appropriate cell passage and growth factor for genetic manipulation [80].

In comparison with single Bac-CB transduction, co-transduction of P3 rabbit chondrocytes with Bac-CB and Bac-CT (BMP-2 and TGF-β1 co-expression) synergistically enhances the expression of aggrecan and collagen IIB (the splice variant form expressed in differentiated chondrocytes) and elevates the deposition of matrix molecules and leads to emergence of chondrocyte-specific lacunae [86]. These data demonstrate that baculovirus-mediated co-expression of growth factor cocktails mounts synergistic effects to coordinate the re-differentiation process of partially de-differentiated P3 chondrocytes. However, Bac-CB and Bac-CT co-transduction also upregulates the de-differentiation marker collagen I and hypertrophy marker collagen X [86].

Since cartilaginous constructs in static cultures often contain a hypoxic necrotic central region and dense layers of cells in the construct periphery [87] and Bac-CB transduction alone is insufficient to support uniform 3D cartilage growth in the static culture [88], Bac-CB-transduced P3 rabbit articular chondrocytes are seeded into poly (L-lactide-*co*-glycolide) (PLGA) scaffolds and cultured in the RSB in order to address the needs for mass transport and mechanical stimuli [88]. Albeit the transient BMP-2 expression, after 3-week culture in the RSB the Bac-CB-transduced constructs grow into cartilage-like tissues with hyaline appearance, uniform cell distribution, enhanced cartilage-specific gene expression and ECM deposition. The GAGs and collagen yield at week 3 are superior to those of cartilaginous tissues cultured in other reactors for longer periods of time [89–92].

To examine how the in vitro culture time influences the maturity of the engineered cartilaginous constructs and how this parameter influences the in vivo repair of osteochondral defects in rabbits, in a subsequent study the de-differentiated P3 chondrocytes are transduced ex vivo with Bac-CB, seeded to PLGA scaffolds and cultured statically (in culture dishes) for 1 day (Bac-w0 group) or in the RSB for 1 week (Bac-w1 group) or 3 weeks (Bac-w3 group) [60]. Bac-CB transduction and increasing culture time in the RSB generate more mature cartilaginous constructs as

**Fig. 4.1** (continued) The expression level is high in a cell passage-dependent manner. For all virus doses (multiplicity of infection) tested, P1 cells express significantly ($p<0.05$) higher levels of TGF-β1 and IGF-1 than P3 and P5 cells, whereas P3 cells express significantly more BMP-2 than P1 and P5 cells. The expression is also dose-dependent, with the highest concentration attained at MOI 100 for TGF-β1 (95 ± 18 ng/ml for P1 cells), at MOI 25 for IGF-1 (82 ± 7 ng/ml for P1 cells) and at MOI 75 for BMP-2 (1,047 ± 76 ng/ml for P3 cells), respectively (With permission of Sung et al. [80])

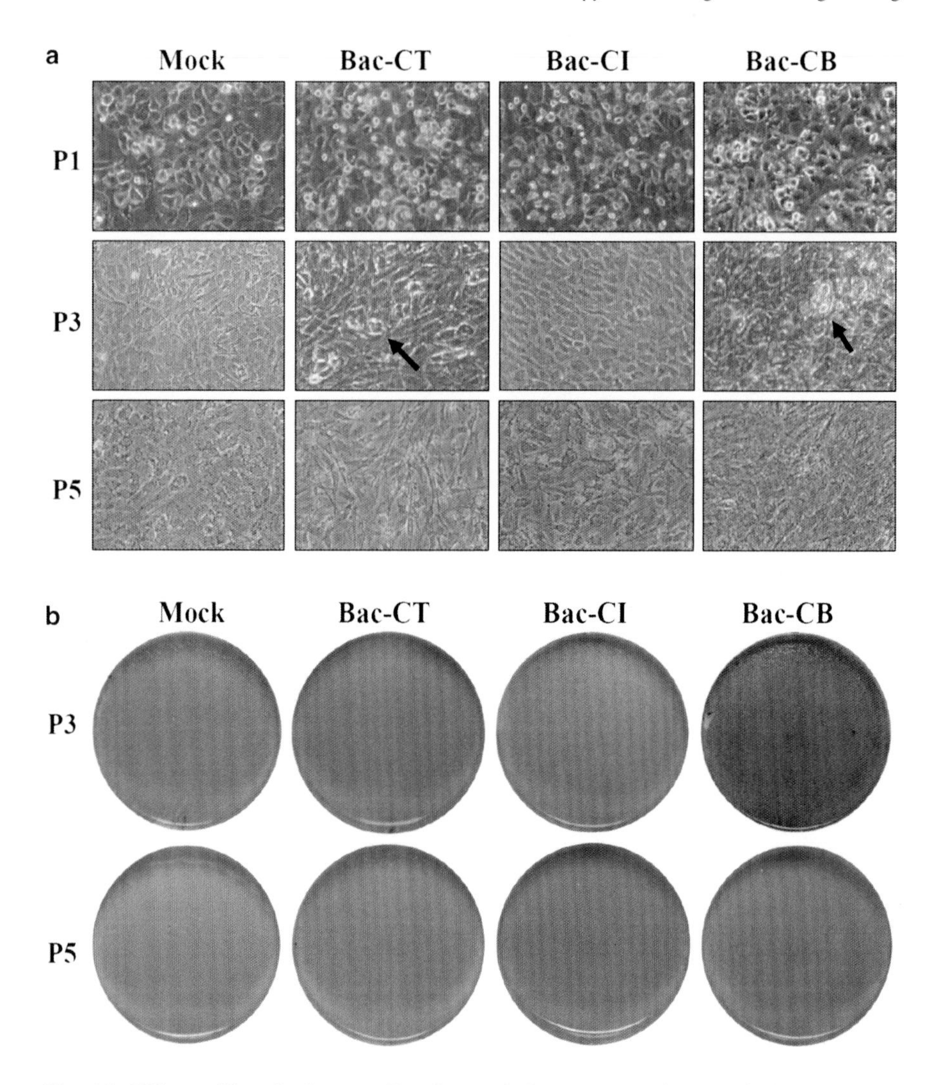


**Fig. 4.2 Effects of baculovirus-mediated growth factor expression**. (**a**) Cell morphology. (**b**) GAGs deposition. The chondrocytes of different passages (P1, P3 and P5) are either mock-transduced (Mock) or transduced by Bac-CT (MOI 100), Bac-CI (MOI 25) or Bac-CB (MOI 75). As shown in the micrographs (200X) captured at 5 days post-transduction (**a**), the majority of mock-transduced cells progressively de-differentiate upon subculture as judged from the change of cell morphology from chondrocyte-like (P1) to spindle-shaped (P3 and P5). Transduction of P3 cells with Bac-CT or Bac-CB result in the emergence of cell nodules (as indicated by *arrows*) and a markedly higher percentage of round or polygonal cells, which indicates the restoration of de-differentiated phenotype. Bac-CI transduction of P3 cells leads to less prominent restoration of chondrocyte phenotype. All the P5 cells, regardless of being transduced by which virus, appear rather de-differentiated. The toluidine blue staining analyzed at day 5 shows that only Bac-CB transduction of P3 cells results in abundant accumulation of GAGs. Bac-CT and Bac-CI only lead to scarce GAGs deposition (With permission of Sung et al. [80])

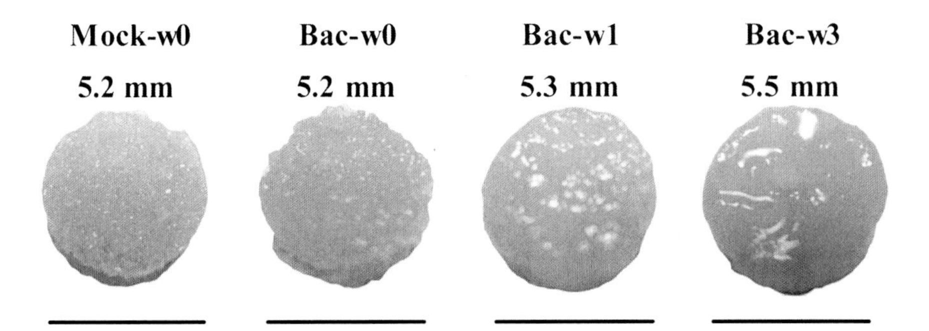


**Fig. 4.3 Effects of in vitro culture time on the engineered cartilage size and appearance.** P3 rabbit articular chondrocytes are transduced with Bac-CB at MOI 75 and seeded into PLGA scaffolds. The transduced constructs cultured statically for 1 day are designated Bac-w0. The transduced constructs cultured in the RSB for 1 or 3 weeks are designated Bac-w1 or Bac-w3. In parallel, the mock-transduced cells are seeded to PLGA scaffolds, cultured in the dish for 1 day and serve as the control (Mock-w0). The diameter of the blank scaffold is ≈5.2 mm. Bar = 5 mm. Mock-w0 and Bac-w0 constructs barely grow in size and deposit nearly no ECM, while the diameter increased to ≈5.3 mm and ≈5.5 mm for the constructs cultured in the RSB for 1 (Bac-w1) and 3 (Bac-w3) weeks, respectively. Accumulation of more hyaline material on the transduced constructs over culture time is clearly visible (With permission of Chen et al. [60])

judged from the increased size/ECM accumulation (Fig. 4.3) as well as ECM composition and mechanical properties (Fig. 4.4).

Eight weeks after implantation into the osteochondral defects at the patellar grooves of New Zealand White (NZW) rabbits, Bac-w0 constructs result in augmented, yet incomplete, repair (Fig. 4.5). The use of Bac-w0 constructs mimics the commonly employed strategies whereby the cells are genetically modified, embedded into scaffolds and implanted immediately [57, 58, 94, 95]. This result suggests that implantation of mechanically immature constructs may lead to tissue deformation that compromises subsequent tissue integration and remodeling. Bac-w1 constructs yield neocartilage layers rich in collagen II and GAGs, but the integration between the graft and host cartilages is not complete. In contrast, Bac-w3 constructs give rise to the regeneration of hyaline cartilages as characterized by cartilage-like appearance, improved integration, chondrocytes clustered in lacunae, smooth and homogeneous matrix rich in collagen II and GAGs but deficient in collagen I (Fig. 4.5).

### *4.3.2 Gene Transfer to Chondroprogenitor Cells*

Despite the success of using articular chondrocytes, supply of autologous cartilage is severely limited and this requires two surgical procedures (isolation of autologous cells and re-implantation), and the cells tend to de-differentiate upon serial cell

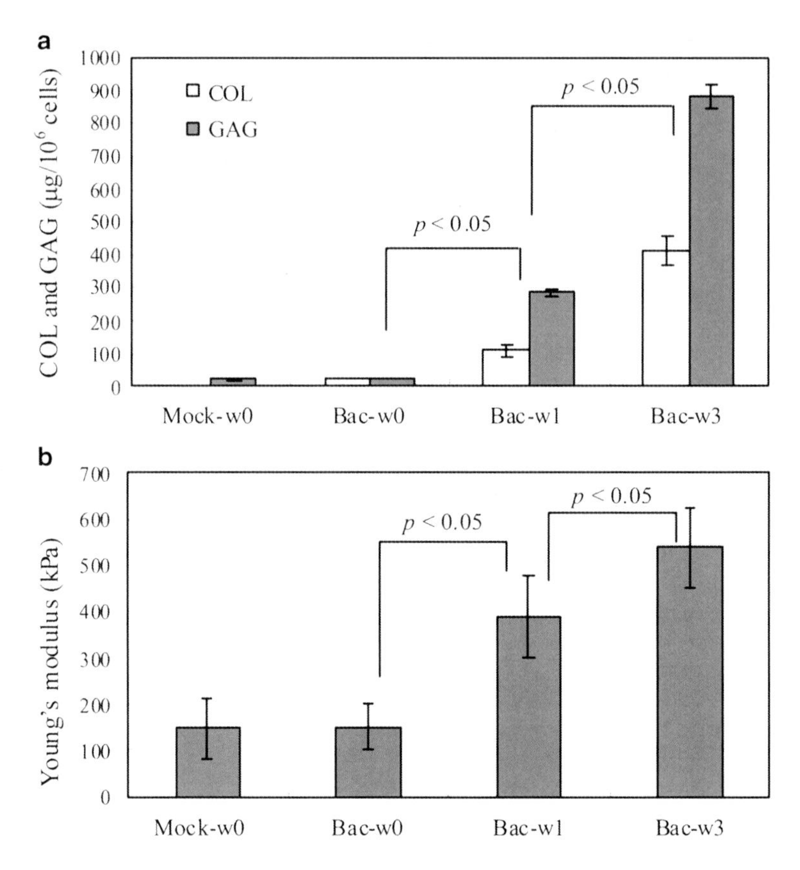


**Fig. 4.4 Effects of in vitro culture time on the ECM accumulation and mechanical properties**. (**a**) Total specific yield of collagen and GAG. (**b**) Young's moduli. The biochemical analyses of total collagen and GAGs (**a**) confirm that scarce ECM is deposited in the Mock-w0 and Bac-w0 constructs, whereas the specific total yield of collagen and GAGs dramatically increased for Bac-w1 and Bac-w3 constructs. Concurrent with the ECM accumulation, the Young's moduli of Bac-w1 and Bac-w3 constructs increase to $388 \pm 89$ kPa and $537 \pm 85$ kPa, respectively, which corresponded to $\approx 50$ % and $\approx 70$ % strength of the native rabbit articular cartilages ($\approx 800$ kPa, [93]) (With permission of Chen et al. [60])

expansion. Therefore, other chondroprogenitor cells have been tested as the target cells for gene transfer. Synoviocytes are the cells in the synovial membrane and can be co-transduced with adenovirus expressing IGF-1 and IL-1Ra in culture, which results in enhanced cartilage matrix synthesis and returns cartilage proteoglycan content to normal levels even when the cells are exposed to IL-1 [22]. Autologous synovial fibroblasts are also transduced with a retrovirus expressing IL-1Ra. Injection of the genetically modified cells into human joints alleviates disease, thus justifying the use of genetically engineered cells for the treatment of arthritis and related disorders [62].

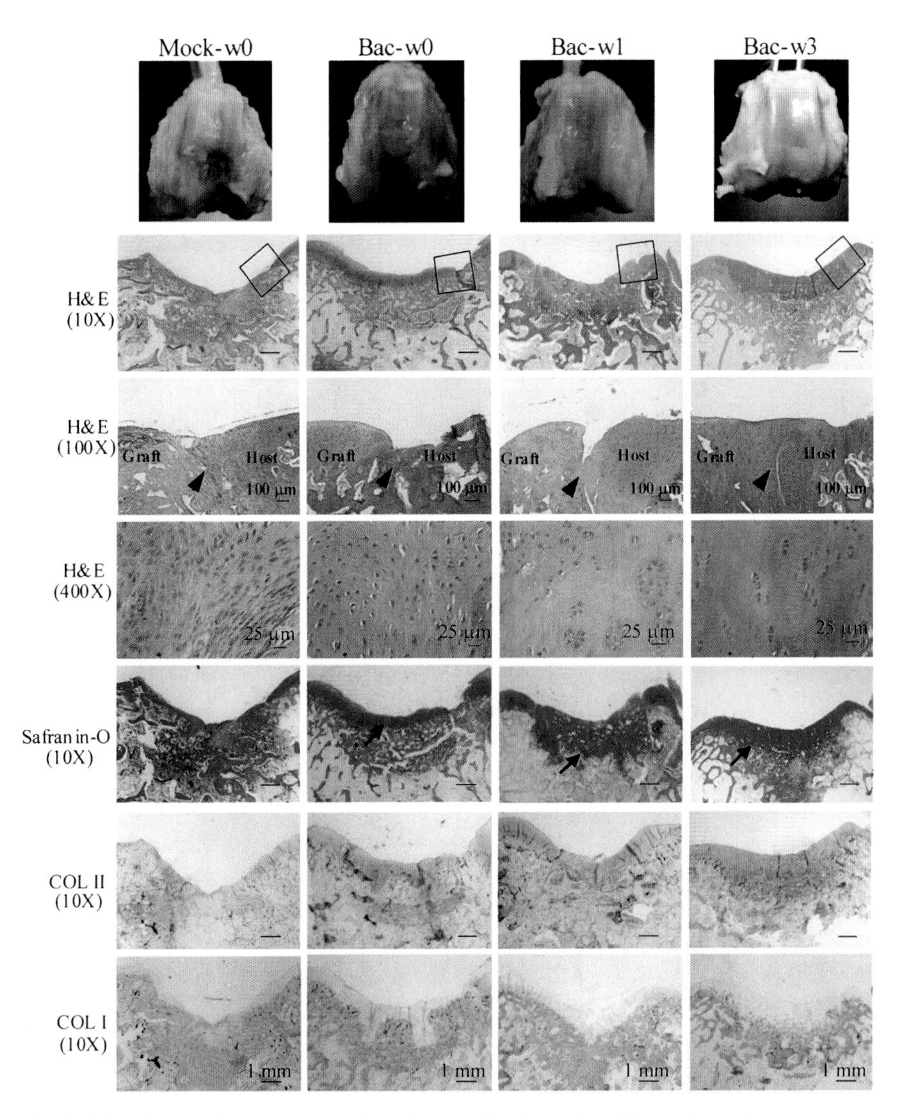


**Fig. 4.5 In vivo cartilage repair at 8 weeks post-implantation**. The engineered constructs as prepared in Fig. 4.3 (Bac-w0, Bac-w1 or Bac-w3) are implanted into the full-thickness defect (5 mm in diameter and 3 mm in depth) in the patella groove of NZW rabbits. The rabbits are sacrificed at 8 weeks post-implantation. The grafts are sectioned for H&E staining, Safranin-O staining (for GAGs) and immunohistochemical staining for collagen II and I (With permission of Chen et al. [60])

Perichondral/periosteal cells are also capable of chondrogenesis when stimulated with chondroinductive factors [96], thus periosteal stem cells are transfected with genes encoding BMP-7 and sonic hedgehog (Shh), seeded to polymer scaffolds and implanted to full-thickness osteochondral defects of NZW rabbits [63]. Both groups

(BMP-7 and Shh) significantly enhance the quality of the repair tissue, resulting in a much smoother surface and more hyaline-appearing cartilage. Notably, the cells expressing Shh regenerate the cartilage in rabbits more efficiently than BMP-7, with an increase in proteoglycan content and integration into the surrounding tissue [63]. In another study, mesenchymal cells isolated from rib perichondrium are co-transduced ex vivo with adenoviral vectors expressing BMP-2 or IGF-1. The cells are suspended in fibrin glue and applied to mechanically induced partial-thickness cartilage lesions in the patellar groove of the rat femur [64]. Transplanted cells are capable of attaching to the wounded articular cartilage and are not displaced from the lesions by joint movement. Engineered cells expressing both BMP-2 and IGF-1 result in the repair of cartilage with hyaline morphology and matrix deficient in collagen I but rich in collagen II and proteoglycan [64]. Untransduced cells either fail to fill up the defects or form fibrocartilage mainly composed of collagen I. Of note, excessive cells are partially dislocated to the joint margins, leading to osteophyte formation if cells transduced with the adenovirus expressing BMP-2 are used. These adverse effects, however, are not observed with the cells transduced with adenovirus expressing IGF-1 [64].

### *4.3.3 Gene Transfer to Mesenchymal Stem Cells (MSCs)*

As mentioned above, MSCs are capable of differentiation into chondrocytes under appropriate environmental cues [97], rendering MSCs a promising cell therapy platform in regenerative medicine [97, 98]. Furthermore, embryonic mesenchymal cell line (C3H10T1/2) is transduced with a retrovirus vector expressing BMP-2. Micromass culture (which stimulates chondrogenesis of MSCs) of the cells and BMP-2 expression is able to induce the chondrogenic differentiation of transduced cells [99].

The first demonstration of MSCs in cartilage repair is performed in a NZW rabbit model with full-thickness lesions and filled with collagen sponges saturated with MSCs. The MSCs differentiate into chondrocytes that secrete cartilaginous matrix [100]. However, this approach results in a discontinuity between the host tissue and the new tissue, and a progressive thinning of the repaired tissue [101].

To orchestrate the differentiation of MSCs, human MSCs are cultured in chondroinduction medium containing ITS+ (insulin-transferrin-selenium) and 10 ng/ml TGF-β3 for up to 7 weeks [102]. The differentiation cascade initiated in MSCs is primarily characterized by sequential upregulation of cartilage genes. Premature induction of hypertrophy-related molecules such as collagen X and metalloprotease 13 (MMP13) occurs before the production of collagen II and is followed by upregulation of alkaline phosphatase (ALP), thus MSC pellets mineralize, in spite of persisting proteoglycan and collagen II content. After transplantation into ectopic sites in SCID mice, MSC pellets undergo endochondral ossification rather than adopting a stable chondrogenic phenotype [102].

MSCs can be genetically engineered with different vectors, including plasmid [103], retrovirus [104], adenovirus [105] and lentivirus [106]. Similarly, baculovirus can efficiently transduce human MSCs [107] and chondrogenic progenitors originating from MSCs [108]. By genetic modification, the expressed therapeutic proteins can promote or modulate the cellular differentiation and accelerate tissue/organ regeneration in vitro or in vivo [105, 106], rendering genetically modified MSCs a promising platform for cartilage gene therapy.

How well the bone marrow-derived MSCs (BMSCs) in aggregate culture can respond to adenovirus-expressed growth factors is compared by transducing the cells with adenovirus expressing TGF-β1, BMP-2 or IGF-1 [109], which reveals that adenovirus-mediated expression of TGF-β1 and BMP-2, but not IGF-1, induce chondrogenesis of MSCs [109]. The chondrogenesis correlates with the protein expression level and duration, and is strongest in aggregates expressing the transgene product at the levels of 10–100 ng/ml. However, chondrogenesis is inhibited in aggregates expressing >100 ng/ml TGF-β1 or BMP-2 [109]. These data highlight the significance of optimal conditions (e.g. viral load and gene expression level) to induce chondrogenesis. In a follow-up study [110], the combinatory effects of these growth factors on chondrogenesis are examined. BMSCs are transduced with each adenoviral vector individually, or in combination, and are cultured in aggregated form for 3 weeks in a defined serum-free medium. Levels of transgenes products in the medium are initially high and decline thereafter. When compared with expression of single gene products, co-expression of IGF-1 and TGF-β1, BMP-2 at low doses results in larger aggregates, higher production levels of GAGs, proteoglycans, collagen II/X, and greater expression of cartilage-specific marker genes. Gene-induced chondrogenesis of MSCs using multiple genes that act synergistically may enable the administration of reduced viral doses in vivo and could be advantageous for the development of cell-based therapies for cartilage repair [110].

Adult human BMSCs are also transduced with AAV expressing green fluorescent protein (GFP) or TGF-β1 and studied in pellet cultures or implanted into osteochondral defects of athymic rats [6]. In pellet culture, GFP expression is visualized through 21 days in vitro. In vivo GFP transgene expression is observed by in situ fluorescent surface imaging in 100 % of GFP implanted defects at week 2, 67 % at week 8 and 17 % at week 12. Improved cartilage repair is observed in osteochondral defects implanted with BMSCs transduced with AAV expressing TGF-β1 at week 12 [6].

Furthermore, BMSCs are transduced with the adenovirus vector expressing Nell-1, seeded into PLGA scaffolds [65] and implanted into osteochondral defects in the central part of the mandibular condyle in adult goats. The Nell-1-modified BMSCs/PLGA constructs result in vigorous and rapid repair leading to regeneration of fibrocartilage at week 6 and to complete repair of native articular cartilage and subchondral bone at week 24. The BMSCs/PLGA group also completely repairs the defect with fibrocartilage at week 24, but the cartilage in the BMSCs/PLGA group is less well-organized than the Nell-1-modified BMSCs/PLGA. The osteochondral defects in the PLGA and empty defect groups are poorly repaired, and no cartilage in the empty defect group or only small portion of cartilage in the PLGA group is found [65]. Therefore, Nell-1-modified BMSCs/PLGA composite can rapidly repair large osteochondral defect in the mandibular condyle with regeneration of native fibrocartilage and subchondral bone [65].

SOX family genes play crucial roles in regulating the chondrogenesis cascades. SOX-9 gene can be delivered into mouse BMSCs via lipofection, thereby enhancing the chondrogenesis of these cells in high density micromass culture [66]. When the transfected MSCs are loaded into the diffusion chamber and transplanted into the athymic mice, massive cartilage-like tissues develop in the chamber 4 weeks after transplantation [66]. Rabbit BMSCs can also be efficiently transduced with an adenoviral vector expressing SOX-9, which induces chondrogenesis both in monolayer and on polyglycolic acid (PGA) scaffold effectively [26]. After implantation into the full-thickness cartilage defects in rabbits, the constructs comprising PGA scaffold and SOX-9-expressing BMSCs result in more neocartilages, hyaline cartilage-specific ECM and greater expression of chondrogenic marker genes than controls [26].

Kim and Im [111] hypothesize that SOX trio genes (SOX-5, SOX-6, and SOX-9) have lower levels of expression during the chondrogenic differentiation of BMSCs compared with chondrocytes, and electroporation of SOX trio genes can promote chondrogenesis of human BMSCs. Indeed, in the in vitro pellet culture without TGF-β1, untransfected BMSCs have a lower level of SOX trio gene and protein expression than chondrocytes [111]. However, the level of SOX-9 gene expression increases in BMSCs when treated with TGF-β1. Co-transfection with SOX trio genes significantly increases the GAG level, collagen IIA1 gene and protein and decreases collagen XA1 protein in BMSCs. Therefore, electroporation-mediated SOX trio gene delivery enhances chondrogenesis and suppresses hypertrophy of human BMSCs [111]. The SOX trio genes are also delivered into human BMSCs via a polyplex consisting of poly(ethyleneimine) (PEI) and PLGA nanoparticles [112]. Such polyplex system considerably improves the transfection efficiencies for human BMSCs. SOX trio genes complexed with PEI-modified PLGA nanoparticles also leads to a dramatic increase in the chondrogenesis of human BMSCs in in vitro culture systems [112].

### *4.3.4 Gene Transfer to ASCs*

Similar to BMSCs, ASCs are multipotent stem cells capable of chondrogenesis and have gained growing popularity for cartilage regeneration [113] as they can be easily obtained in large quantities from liposuction and commit chondrogenesis when cultured in chondrogenic medium containing TGF-β1, TGF-β2, TGF-β3, BMP-2 or BMP-6 [21, 114, 115].

Human ASCs can be transduced with an adenovirus expressing TGF-β2, and pre-differentiated in vitro by culturing in 12 well plates using chondrogenic medium devoid of growth factors [8]. The pre-differentiated ASCs are seeded to different scaffolds and implanted into subcutaneous pockets on the dorsum of nude mice. At 4 and 12 weeks post-implantation, cartilage-like tissue formation is only found in the alginate gel and PLGA/alginate groups, but in the PLGA group fibrous tissues and angiogenesis are observed. These findings demonstrate that adenovirus-mediated TGF-β2 expression can induce ASCs differentiation into chondrogenic lineage in vitro [8]. However, this pre-differentiation does not guarantee ectopic cartilage formation in vivo unless appropriate 3D scaffolds are used as the cell carriers.

Chondrogenesis of ASCs can be stimulated by transfection with a plasmid encoding BMP-6 followed by encapsulation within alginate beads and culture in chondrogenic medium containing synthetic glucocorticoid dexamethasone (DEX) or the combination of epidermal growth factor (EGF), FGF-2, and TGF-β1 [116]. BMP-6 overexpression alone induces a moderate chondrogenic response, yet the inclusion of other growth factors in the medium promotes robust collagen II expression. However, the growth factor combination also increases the deposition of collagen I and X, indicating the induction of a hypertrophic chondrocyte phenotype. Early gene expression data indicates that DEX is synergistic with BMP-6 for chondrogenesis, but DEX reduces GAG accumulation at day 28. These results suggest that chondrogenic differentiation of ASCs depends on complex interactions among various growth factors and media supplements, as well as the concentration and duration of growth factor exposure [116].

In addition, Im and co-workers have delivered the SOX trio (SOX-5, -6, and -9) genes into ASCs by retrovirus [27] and by a chondrogenic scaffold system in which plasmid DNA encoding SOX trio genes is incorporated into a PLGA scaffold and slowly released to transfect ASCs seeded in the scaffold [117]. The ASCs co-transduced with retrovirus are embedded in fibrin gel and implanted into the osteochondral defect created in the patellar groove of the distal femur, and also injected into the knee joints of rats with surgically-induced osteoarthritis [27]. Co-transduction with SOX trio significantly increases GAG contents as well as collagen II gene and protein expression. ASCs co-transduced with SOX trio also significantly promote the in vivo cartilage healing in the osteochondral defect model, and prevent the progression of degenerative changes in surgically-induced osteoarthritis.

Additionally, we have employed a hybrid baculovirus system that exploits FLPo/Frt-mediated transgene recombination and episomal minicircle formation to genetically engineer rabbit ASCs (see Chap. 2 and [67]). Three recombinant baculoviruses are constructed: one expressing FLPo while the other two baculoviruses harbor transgenes encoding TGF-β3 and BMP-6, respectively. The TGF-β3/BMP-6 gene cassettes are flanked by Frt sequences, so that after co-transduction of rabbit ASCs with the three baculoviruses FLPo recognizes the Frt sequences and mediates the recombination and formation of minicircles. The hybrid baculovirus system confers prolonged and robust TGF-β3/BMP-6 expression in ASCs seeded into porous scaffolds. Two week culture in vitro augments ASCs chondrogenesis and suppresses osteogenesis/hypertrophy, leading to the formation of cartilaginous constructs (Fig. 4.6a) with improved maturity and mechanical properties. Implantation of the resultant engineered constructs into the load-bearing, full-thickness articular cartilage defects in NZW rabbits leads to progressive defect healing (Fig. 4.6b). Twelve weeks after implantation into full-thickness articular cartilage defects in rabbits, these engineered constructs regenerate neocartilages that resemble native hyaline cartilages in gross appearance (Fig. 4.6c), cell morphology (Fig. 4.6d), matrix composition (Fig. 4.7) and mechanical properties [67]. In particular, the neocartilages display a native cartilage-characteristic zonal structure (superficial, middle, deep and

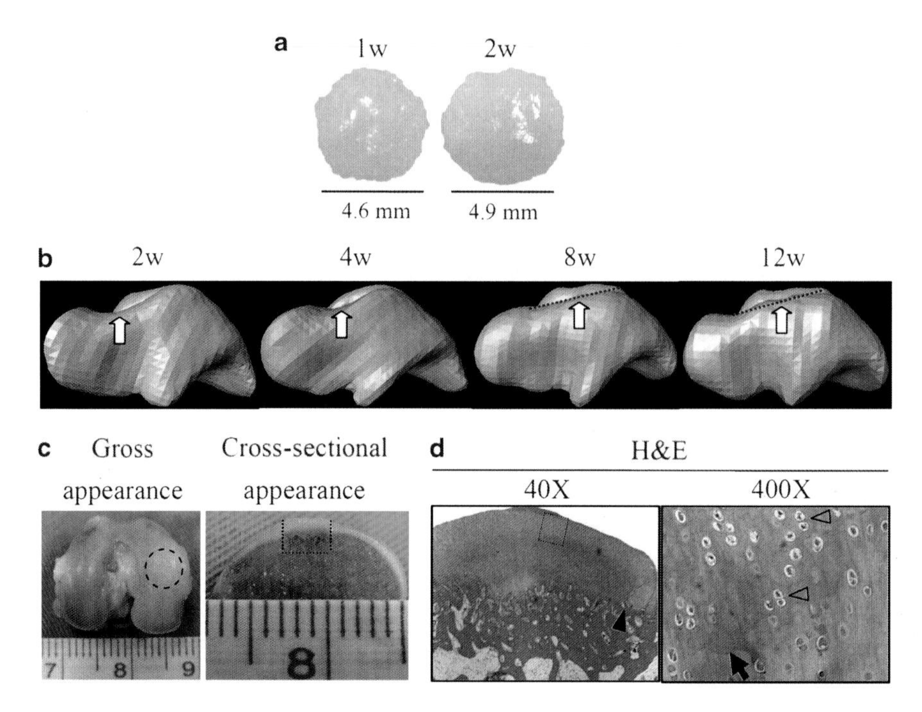


**Fig. 4.6 Preparation of hybrid baculovirus-engineered constructs and use of constructs for in vivo cartilage repair**. (**a**) Gross appearance of the engineered cartilage after 1 (1w) and 2 (2w) week culture. (**b**) Progressive cartilage repair at different weeks post-transplantation. (**c**) Gross appearance of the knee joints removed from the rabbits at 12 weeks post-transplantation. (**d**) H&E staining of the repaired cartilages. Rabbit ASCs are co-transduced with Bac-FLPo, Bac-FCT3W and Bac-FCB6W, seeded to scaffolds (≈4 mm in diameter, ≈3 mm in thickness, $2 \times 10^6$ cells/scaffold) and cultured in 12-well plates using chondrogenic medium under hypoxic conditions (5 % $O_2$, 5 % $CO_2$ and 90 % $N_2$). After 2-week in vitro culture, the engineered constructs are transplanted into the full-thickness articular cartilage defects in NZW rabbits and the repair is monitored by magnetic resonance imaging (MRI) at different weeks post-transplantation. At week 12, the knees are removed from rabbits for observation and analysis (With permission of Lu et al. [67])

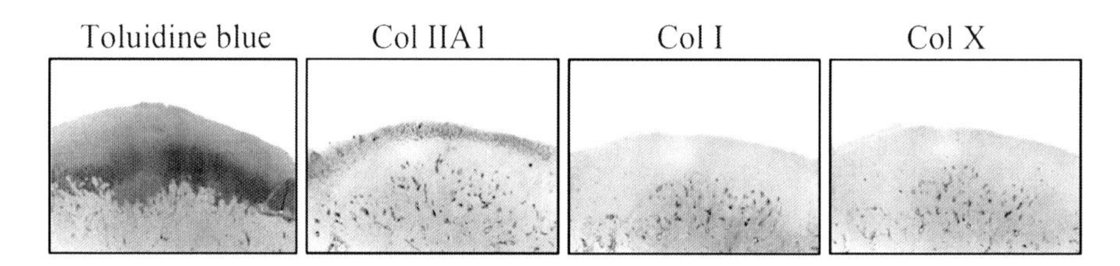


**Fig. 4.7 ECM composition of the regenerated cartilages**. The neocartilages removed at 12 weeks post-transplantation as in Fig. 4.6 are sectioned and subjected to toluidine blue staining and immunohistochemical staining specific for collagen IIA1, collagen I and collagen X (With permission of Lu et al. [67])

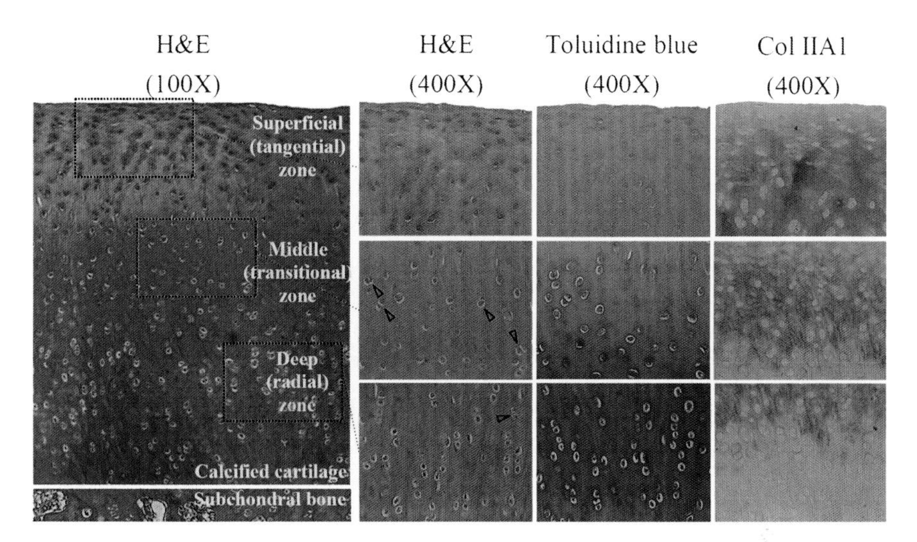


**Fig. 4.8 Zonal structure of the regenerated neocartilage**. The neocartilages removed at 12 weeks post-transplantation as in Fig. 4.6 are sectioned and subjected to analysis. The articular cartilage-specific layered structure (superficial, middle, deep and calcified zones) is observed in the neocartilage. The cell morphology, matrix composition and orientation resemble those of the native cartilage (With permission of Lu et al. [67])

calcified zones) (Fig. 4.8) and are histochemically and biomechanically superior to control groups. Since each zone of the articular cartilages plays important roles in the biochemical and mechanical functions [118], recapitulating such characteristic layers is beneficial for the functionality and long-term stability of the repaired cartilages [119]. Consequently, the neocartilages at week 24 integrate well with the host cartilages without signs of degeneration, successfully repair the defects and undergo neither fibrosis nor ossification, thus proving the long-term stability of the neocartilages [67]. In contrast, ASCs that are engineered with the non-hybrid baculovirus vectors transiently expressing TGF-β3/BMP-6 undergo osteogenesis/hypertrophy and result in the formation of inferior cartilaginous constructs, which after implantation regenerate fibrocartilages [67]. These data underscore the crucial role of TGF-β3/BMP-6 expression level and duration in ASCs in the cell differentiation, constructs properties and in vivo repair. The hybrid baculovirus-engineered ASCs that persistently express TGF-β3/BMP-6 improve the chondrogenesis, in vitro cartilaginous constructs production and in vivo hyaline cartilage regeneration.

### *4.3.5 Gene Transfer via Gene Activated Matrix (GAM)*

Analogous to gene transfer for bone engineering, gene transfer for cartilage engineering can be mediated via GAM, yet in the context of cartilage repair the GAM is often seeded with cells for ex vivo gene delivery. Since subchondral bone underlies the cartilage, attempts have been made to fabricate bi-layered scaffolds in which one layer stimulates the cartilage formation while the other layer stimulates subchondral bone formation. Using this concept, a bi-layered gene activated osteochondral scaffold has been designed [120]. One layer of this GAM is the chitosan-gelatin scaffold encompassing the plasmid encoding TGF-β1 and the second layer is the hydroxyapatite/chitosan-gelatin scaffold encompassing the plasmid coding for BMP-2 for inducing osteogenesis [120]. BMSCs are seeded to each layer of the bi-layered gene activated osteochondral scaffold, which results in significant cell proliferation, high expression of TGF-β1 and BMP-2. The spatially controlled and localized gene delivery system in the bi-layered integrated scaffolds can induce the MSCs in different layers to differentiate into chondrocytes and osteoblasts in vitro, respectively, and simultaneously support the articular cartilage and subchondral bone regeneration in the rabbit knee osteochondral defect model. This study demonstrates the feasibility of complex tissue regeneration through the combination of biomimetic and multi-phasic scaffold design and localized gene delivery system [120].

In another study, a composite GAM consisting of BMSCs, plasmid DNA encoding TGF-β1, fibrin gel and PLGA sponge is designed and employed to repair articular cartilage defects [103]. Since transfection efficiency of BMSCs is generally low, the transfection efficiency is improved by using a cationized chitosan derivative N,N,N-trimethyl chitosan chloride (TMC) as a carrier. The TMC/DNA complexes result in a transfection efficiency for BMSCs of 9 % and enable heterogeneous TGF-β1 expression in a 10-day culture period in vitro [103]. After implantation of the constructs into full-thickness cartilage defects of NZW rabbit joints, in vivo TGF-β1 expression is detectable at week 4 although its level decreases with time. At 12 weeks post-implantation, the cartilage defects are successfully repaired by the GAM constructs, and the neocartilage integrates well with its surrounding tissue and subchondral bone. Immunohistochemical and GAGs staining confirm the similar amount and distribution of collagen II and GAGs in the regenerated cartilage as that of hyaline cartilage [103]. In contrast, only part of the defect is repaired by the constructs lacking TGF-β1-encoding plasmid, and only fibrous tissue is found in the defects filled with the GAM construct lacking BMSCs. Therefore, combination of GAM with BMSCs holds promise to restore cartilage defects [103].

### *4.3.6 Comparison of Cell Sources*

With the availability of different cell sources, it is of interest to compare the chondrogenic potential of different cells. To this end, mesenchymal cells are isolated from perichondrium/periosteum, bone marrow or fat of adult rats [68]. The cells are transduced with an adenovirus expressing BMP-2 or stimulated with recombinant BMP-2 [68]. Stimulation with BMP-2 or adenovirus leads to up-regulation of cartilage-specific gene expression in all three cell populations studied, yet the effects are more rapid and prominent in the perichondrial/periosteal cells. The cells transduced with adenovirus are transplanted into partial-thickness cartilage lesions in the patellar groove of the rat femur, after which transduced perichondrial/periosteal cells produce a proteoglycan-rich, collagen II-positive matrix with only faint staining for collagen I [68]. The repair tissue originating from transduced bone marrow stem cells shows less intense collagen II staining, but a relatively proteoglycan-rich matrix and weakly positive for collagen I. Transgene-activated fat stromal cells form rather fibrous tissue mainly composed of collagen I. Therefore, perichondrium/periosteum-derived cells and bone marrow-derived mesenchymal cells seem superior to cells isolated from fat with respect to forming hyaline cartilaginous tissue [68].

In another study, Yang et al. have transfected BMSCs, ASCs and de-differentiated chondrocytes with SOX trio genes, encapsulated the cells in the fibrin hydrogel and grown the cells under pellet culture conditions [121]. Chondrogenic genes and proteins are more highly expressed in SOX trio-expressing cells than in untransfected cells. In addition, not only specific genes and proteins, but cartilage-forming tissues are observed in nude mice transplanted with SOX trio-expressing BMSCs, ASCs, and de-differentiated chondrocytes. Both in vitro and in vivo analyses reveal that cells transfected with the SOX trio genes successfully differentiate into mature chondrocytes and could be used for the reconstruction of hyaline articular cartilage. However, no significant differences in chondrogenesis are found between the three cell sources [121].

## References


1. Wei Y, Hu Y, Lv R, Li D (2006) Regulation of adipose-derived adult stem cells differentiating into chondrocytes with the use of rhBMP-2. Cytotherapy 8:570–579
2. Miljkovic N, Cooper G, Marra K (2008) Chondrogenesis, bone morphogenetic protein-4 and mesenchymal stem cells. Osteoarthritis Cartilage 16:1121–1130
3. Kuroda R, Usas A, Kubo S, Corsi K, Peng H, Rose T et al (2006) Cartilage repair using bone morphogenetic protein 4 and muscle-derived stem cells. Arthritis Rheum 54:433–442
4. Kemmis CM, Vahdati A, Weiss HE, Wagner DR (2010) Bone morphogenetic protein 6 drives both osteogenesis and chondrogenesis in murine adipose-derived mesenchymal cells depending on culture conditions. Biochem Biophys Res Commun 401:20–25
5. Vukicevic S, Grgurevic L (2009) BMP-6 and mesenchymal stem cell differentiation. Cytokine Growth Factor Rev 20:441–448
6. Pagnotto MR, Wang Z, Karpie JC, Ferretti M, Xiao X, Chu CR (2007) Adeno-associated viral gene transfer of transforming growth factor-β1 to human mesenchymal stem cells improves cartilage repair. Gene Ther 14:804–813
7. Han Y, Wei Y, Wang S, Song Y (2010) Cartilage regeneration using adipose-derived stem cells and the controlled-released hybrid microspheres. Joint Bone Spine 77:27–31
8. Jin X, Sun Y, Zhang K, Wang J, Shi T, Ju X et al (2007) Ectopic neocartilage formation from predifferentiated human adipose derived stem cells induced by adenoviral-mediated transfer of hTGF beta2. Biomaterials 28:2994–3003
9. Bouffi C, Thomas O, Bony C, Giteau A, Venier-Julienne M-C, Jorgensen C et al (2010) The role of pharmacologically active microcarriers releasing TGF-β3 in cartilage formation in vivo by mesenchymal stem cells. Biomaterials 31:6485–6493
10. Hennig T, Lorenz H, Thiel A, Goetzke K, Dickhut A, Geiger F et al (2007) Reduced chondrogenic potential of adipose tissue derived stromal cells correlates with an altered TGFbeta receptor and BMP profile and is overcome by BMP-6. J Cell Physiol 211:682–691

11. Feng G, Wan Y, Balian G, Laurencin CT, Li X (2008) Adenovirus-mediated expression of growth and differentiation factor-5 promotes chondrogenesis of adipose stem cells. Growth Factors 26:132–142
12. Fan H, Tao H, Wu Y, Hu Y, Yan Y, Luo Z (2010) TGF-beta3 immobilized PLGA-gelatin/chondroitin sulfate/hyaluronic acid hybrid scaffold for cartilage regeneration. J Biomed Mater Res A 95:982–992
13. Barry F, Boynton RE, Liu B, Murphy JM (2001) Chondrogenic differentiation of mesenchymal stem cells from bone marrow: differentiation-dependent gene expression of matrix components. Exp Cell Res 268:189–200
14. Estes BT, Wu AW, Guilak F (2006) Potent induction of chondrocytic differentiation of human adipose-derived adult stem cells by bone morphogenetic protein 6. Arthritis Rheum 54:1222–1232
15. Moioli EK, Hong L, Mao JJ (2007) Inhibition of osteogenic differentiation of human mesenchymal stem cells. Wound Repair Regen 15:413–421
16. Tang QO, Shakib K, Heliotis M, Tsiridis E, Mantalaris A, Ripamonti U (2009) TGF-beta3: a potential biological therapy for enhancing chondrogenesis. Expert Opin Biol Ther 9:689–701
17. Mehlhorn AT, Niemeyer P, Kaschte K, Muller L, Finkenzeller G, Hartl D et al (2007) Differential effects of BMP-2 and TGF-beta1 on chondrogenic differentiation of adipose derived stem cells. Cell Prolif 40:809–823
18. Shen B, Wei A, Tao H, Diwan AD, Ma DD (2009) BMP-2 enhances TGF-beta3-mediated chondrogenic differentiation of human bone marrow multipotent mesenchymal stromal cells in alginate bead culture. Tissue Eng Part A 15:1311–1320
19. Lim SM, Oh SH, Lee HH, Yuk SH, Im GI, Lee JH (2010) Dual growth factor-releasing nanoparticle/hydrogel system for cartilage tissue engineering. J Mater Sci Mater Med 21:2593–2600
20. Otsuki S, Hanson SR, Miyaki S, Grogan SP, Kinoshita M, Asahara H et al (2010) Extracellular sulfatases support cartilage homeostasis by regulating BMP and FGF signaling pathways. Proc Natl Acad Sci U S A 107:10202–10207
21. Puetzer JL, Petitte JN, Loboa EG (2010) Comparative review of growth factors for induction of three-dimensional in vitro chondrogenesis in human mesenchymal stem cells isolated from bone marrow and adipose tissue. Tissue Eng Part B Rev 16:435–444
22. Nixon AJ, Haupt JL, Frisbie DD, Morisset SS, McIlwraith CW, Robbins PD et al (2005) Gene-mediated restoration of cartilage matrix by combination insulin-like factor-I/interleukin-1 receptor antagonist therapy. Gene Ther 12:177–186
23. Aghaloo T, Cowan CM, Chou YF, Zhang X, Lee H, Miao S et al (2006) Nell-1-induced bone regeneration in calvarial defects. Am J Pathol 169:903–915
24. Lee M, Siu RK, Ting K, Wu BM (2010) Effect of Nell-1 delivery on chondrocyte proliferation and cartilaginous extracellular matrix deposition. Tissue Eng Part A 16:1791–1800
25. Bi WM, Deng JM, Zhang ZP, Behringer RR, de Crombrugghe B (1999) Sox9 is required for cartilage formation. Nat Genet 22:85–89
26. Cao L, Yang F, Liu G, Yu D, Li H, Fan Q et al (2011) The promotion of cartilage defect repair using adenovirus mediated Sox9 gene transfer of rabbit bone marrow mesenchymal stem cells. Biomaterials 32:3910–3920
27. Lee J-M, Im G-I (2012) SOX trio-co-transduced adipose stem cells in fibrin gel to enhance cartilage repair and delay the progression of osteoarthritis in the rat. Biomaterials 33:2016–2024
28. Ghivizzani SC, Lechman ER, Tio C, Mule KM, Chada S, McCormack JE et al (1997) Direct retrovirus-mediated gene transfer to the synovium of the rabbit knee: implications for arthritis gene therapy. Gene Ther 4:977–982
29. Ulrich-Vinther M, Duch MR, Soballe K, O'Keefe RJ, Schwarz EM, Pedersen FS (2004) In vivo gene delivery to articular chondrocytes mediated by an adeno-associated virus vector. J Orthop Res 22:726–734

30. Lechman ER, Jaffurs D, Ghivizzani SC, Gambotto A, Kovesdi I, Mi ZB et al (1999) Direct adenoviral gene transfer of viral IL-10 to rabbit knees with experimental arthritis ameliorates disease in both injected and contralateral control knees. J Immunol 163:2202–2208
31. Oligino T, Ghivizzani SC, Wolfe D, Lechman ER, Krisky D, Mi Z et al (1999) Intra-articular delivery of a herpes simplex virus IL-1Ra gene vector reduces inflammation in a rabbit model of arthritis. Gene Ther 6:1713–1720
32. Pan RY, Chen SL, Xiao X, Liu DW, Peng HJ, Tsao YP (2000) Therapy and prevention of arthritis by recombinant adeno-associated virus vector with delivery of interleukin-1 receptor antagonist. Arthritis Rheum 43:289–297
33. Gouze E, Pawliuk R, Pilapil C, Gouze JN, Fleet C, Palmer GD et al (2002) In vivo gene delivery to synovium by lentiviral vectors. Mol Ther 5:397–404
34. Frisbie DD, Ghivizzani SC, Robbins PD, Evans CH, McIlwraith CW (2002) Treatment of experimental equine osteoarthritis by in vivo delivery of the equine interleukin-1 receptor antagonist gene. Gene Ther 9:12–20
35. Mease PJ, Wei N, Fudman EJ, Kivitz AJ, Schechtman J, Trapp RG et al (2010) Safety, tolerability, and clinical outcomes after intraarticular injection of a recombinant adeno-associated vector containing a tumor necrosis factor antagonist gene: results of a phase 1/2 study. J Rheumatol 37:692–703
36. Mease P, Wei N, Fudman E, Kivitz A, Schechtman J, Trapp R et al (2008) Safety and clinical outcomes after intra-articular administration of a recombinant adeno-associated vector containing a TNF antagonist gene. Arthritis Rheum 58:S433–S434
37. Mi ZB, Ghivizzani SC, Lechman ER, Jaffurs D, Glorioso JC, Evans CH et al (2000) Adenovirus-mediated gene transfer of insulin-like growth factor 1 stimulates proteoglycan synthesis in rabbit joints. Arthritis Rheum 43:2563–2570
38. Mi Z, Ghivizzani SC, Lechman E, Glorioso JC, Evans CH, Robbins PD (2003) Adverse effects of adenovirus-mediated gene transfer of human transforming growth factor beta 1 into rabbit knees. Arthritis Res Ther 5:R132–R139
39. Watanabe S, Imagawa T, Boivin GP, Gao GP, Wilson JM, Hirsch R (2000) Adeno-associated virus mediates long-term gene transfer and delivery of chondroprotective IL-4 to murine synovium. Mol Ther 2:147–152
40. Cucchiarini M, Madry H, Ma C, Thurn T, Zurakowski D, Menger MD et al (2005) Improved tissue repair in articular cartilage defects *in vivo* by rAAV-mediated overexpression of human fibroblast growth factor 2. Mol Ther 12:229–238
41. Cucchiarini M, Thurn T, Weimer A, Kohn D, Terwilliger EF, Madry H (2007) Restoration of the extracellular matrix in human osteoarthritic articular cartilage by overexpression of the transcription factor *SOX9*. Arthritis Rheum 56:158–167
42. Cucchiarini M, Orth P, Madry H (2013) Direct rAAV SOX9 administration for durable articular cartilage repair with delayed terminal differentiation and hypertrophy in vivo. J Mol Med 91:625–636
43. Gouze E, Pawliuk R, Gouze JN, Pilapil C, Fleet C, Palmer GD et al (2003) Lentiviral-mediated gene delivery to synovium: potent intra-articular expression with amplification by inflammation. Mol Ther 7:460–466
44. Moreland LW, Baumgartner SW, Schiff MH, Tindall EA, Fleischmann RM, Weaver AL et al (1997) Treatment of rheumatoid arthritis with a recombinant human tumor necrosis factor receptor (p75)-Fc fusion protein. N Engl J Med 337:141–147
45. Moreland LW (1998) Soluble tumor necrosis factor receptor (p75) fusion protein (ENBREL) as a therapy for rheumatoid arthritis. Rheum Dis Clin North Am 24:579–591
46. Evans CH, Ghivizzani SC, Robbins PD (2008) Arthritis gene therapy's first death. Arthritis Res Ther 10:110
47. Frank KM, Hogarth DK, Miller JL, Mandal S, Mease PJ, Samulski RJ et al (2009) Brief report: investigation of the cause of death in a gene-therapy trial. N Engl J Med 361:161–169
48. Vinatier C, Mrugala D, Jorgensen C, Guicheux J, Noel D (2009) Cartilage engineering: a crucial combination of cells, biomaterials and biofactors. Trends Biotechnol 27:307–314

49. Gelse K, Schneider H (2006) *Ex vivo* gene therapy approaches to cartilage repair. Adv Drug Deliv Rev 58:259–284
50. Lieberman JR, Ghivizzani SC, Evans CH (2002) Gene transfer approaches to the healing of bone and cartilage. Mol Ther 6:141–147
51. Arai Y, Kubo T, Fushiki S, Mazda O, Nakai H, Iwaki Y et al (2000) Gene delivery to human chondrocytes by an adeno associated virus vector. J Rheumatol 27:979–982
52. Ulrich-Vinther M, Maloney MD, Goater JJ, Soballe K, Goldring MB, O'Keefe RJ et al (2002) Light-activated gene transduction enhances adeno-associated virus vector-mediated gene expression in human articular chondrocytes. Arthritis Rheum 46:2095–2104
53. Trippel SB, Ghivizzani SC, Nixon AJ (2004) Gene-based approaches for the repair of articular cartilage. Gene Ther 11:351–359
54. Gelse K, Jiang QJ, Aigner T, Ritter T, Wagner K, Poschl E et al (2001) Fibroblast-mediated delivery of growth factor complementary DNA into mouse joints induces chondrogenesis but avoids the disadvantages of direct viral gene transfer. Arthritis Rheum 44:1943–1953
55. Cottard V, Valvason C, Falgarone G, Lutomski D, Boissier MC, Bessis N (2004) Immune response against gene therapy vectors: influence of synovial fluid on adeno-associated virus mediated gene transfer to chondrocytes. J Clin Immunol 24:162–169
56. Gouze E, Gouze J-N, Palmer GD, Pilapil C, Evans CH, Ghivizzani SC (2007) Transgene persistence and cell turnover in the diarthrodial joint: implications for gene therapy of chronic joint diseases. Mol Ther 15:1114–1120
57. Madry H, Kaul G, Cucchiarini M, Stein U, Zurakowski D, Remberger K et al (2005) Enhanced repair of articular cartilage defects *in vivo* by transplanted chondrocytes overexpressing insulin-like growth factor I (IGF-I). Gene Ther 12:1171–1179
58. Hidaka C, Goodrich LR, Chen CT, Warren RF, Crystal RG, Nixon AJ (2003) Acceleration of cartilage repair by genetically modified chondrocytes overexpressing bone morphogenetic protein-7. J Orthop Res 21:573–583
59. Yokoo N, Saito T, Uesugi M, Kobayashi N, Xin K-Q, Okuda K et al (2005) Repair of articular cartilage defect by autologous transplantation of basic fibroblast growth factor gene-transduced chondrocytes with adeno-associated virus vector. Arthritis Rheum 52:164–170
60. Chen H-C, Chang Y-H, Chuang C-K, Lin C-Y, Sung L-Y, Wang Y-H et al (2009) The repair of osteochondral defects using baculovirus-mediated gene transfer with de-differentiated chondrocytes in bioreactor culture. Biomaterials 30:674–681
61. Zhang XL, Mao ZB, Yu CL (2004) Suppression of early experimental osteoarthritis by gene transfer of interleukin-1 receptor antagonist and interleukin-10. J Orthop Res 22:742–750
62. Evans CH, Robbins PD, Ghivizzani SC, Wasko MC, Tomaino MM, Kang R et al (2005) Gene transfer to human joints: progress toward a gene therapy of arthritis. Proc Natl Acad Sci U S A 102:8698–8703
63. Grande DA, Mason J, Light E, Dines D (2003) Stem cells as platforms for delivery of genes to enhance cartilage repair. J Bone Joint Surg Am 85A(suppl 2):111–116
64. Gelse K, von der Mark K, Aigner T, Park J, Schneider H (2003) Articular cartilage repair by gene therapy using growth factor- producing mesenchymal cells. Arthritis Rheum 48:430–441
65. Zhu S, Zhang B, Man C, Ma Y, Hu J (2011) NEL-like molecule-1-modified bone marrow mesenchymal stem cells/poly lactic-co-glycolic acid composite improves repair of large osteochondral defects in mandibular condyle. Osteoarthritis Cartilage 19:743–750
66. Tsuchiya H, Kitoh H, Sugiura F, Ishiguro N (2003) Chondrogenesis enhanced by overexpression of *sox9* gene in mouse bone marrow-derived mesenchymal stem cells. Biochem Biophys Res Commun 301:338–343
67. Lu C-H, Yeh T-S, Yeh C-L, Fang Y-HD, Sung L-Y, Lin S-Y et al (2014) Regenerating cartilages by engineered ASCs: prolonged TGF-β3/BMP-6 expression improved articular cartilage formation and restored zonal structure. Mol Ther 22:186–195
68. Park J, Gelse K, Frank S, von der Mark K, Aigner T, Schneider H (2006) Transgene-activated mesenchymal cells for articular cartilage repair: a comparison of primary bone marrow-, perichondrium/periosteum- and fat-derived cells. J Gene Med 8:112–125

69. Shuler FD, Georgescu HI, Niyibizi C, Studer RK, Mi Z, Johnstone B et al (2000) Increased matrix synthesis following adenoviral transfer of a transforming growth factor $\beta_1$ gene into articular chondrocytes. J Orthop Res 18:585–592
70. Brower-Toland BD, Saxer RA, Goodrich LR, Mi ZB, Robbins PD, Evans CH et al (2001) Direct adenovirus-mediated insulin-like growth factor I gene transfer enhances transplant chondrocyte function. Hum Gene Ther 12:117–129
71. Obradovic B, Martin I, Padera RF, Treppo S, Freed LE, Vunjak-Novakovic G (2001) Integration of engineered cartilage. J Orthop Res 19:1089–1097
72. Martin I, Wendt D, Heberer M (2004) The role of bioreactors in tissue engineering. Trends Biotechnol 22:80–86
73. Chen H-C, Hu Y-C (2006) Bioreactors for tissue engineering. Biotechnol Lett 28:1415–1423
74. Darling EM, Athanasiou KA (2003) Articular cartilage bioreactors and bioprocesses. Tissue Eng 9:9–26
75. Portner R, Nagel-Heyer S, Goepfert C, Adamietz P, Meenen NM (2005) Bioreactor design for tissue engineering. J Biosci Bioeng 100:235–245
76. Madry H, Padera R, Seidel J, Langer R, Freed LE, Trippel SB et al (2002) Gene transfer of a human insulin-like growth factor I cDNA enhances tissue engineering of cartilage. Hum Gene Ther 13:1621–1630
77. Ho Y-C, Chen H-C, Wang K-C, Hu Y-C (2004) Highly efficient baculovirus-mediated gene transfer into rat chondrocytes. Biotechnol Bioeng 88:643–651
78. Chen H-C, Lee H-P, Sung M-L, Liao C-J, Hu Y-C (2004) A novel rotating-shaft bioreactor for two-phase cultivation of tissue-engineered cartilage. Biotechnol Prog 20:1802–1809
79. Chen H-C, Lee H-P, Ho Y-C, Sung M-L, Hu Y-C (2006) Combination of baculovirus-mediated gene transfer and rotating-shaft bioreactor for cartilage tissue engineering. Biomaterials 27:3154–3162
80. Sung L-Y, Lo W-H, Chiu H-Y, Chen H-C, Chuang C-K, Lee H-P et al (2007) Modulation of chondrocyte phenotype via baculovirus-mediated growth factor expression. Biomaterials 28:3437–3447
81. Smith P, Shuler FD, Georgescu HI, Ghivizzani SC, Johnstone B, Niyibizi C et al (2000) Genetic enhancement of matrix synthesis by articular chondrocytes: comparison of different growth factor genes in the presence and absence of interleukin-1. Arthritis Rheum 43:1156–1164
82. Dinser R, Kreppel F, Zaucke F, Blank C, Paulsson M, Kochanek S et al (2001) Comparison of long-term transgene expression after non-viral and adenoviral gene transfer into primary articular chondrocytes. Histochem Cell Biol 116:69–77
83. Madry H, Cucchiarini M, Terwilliger EF, Trippel SB (2003) Recombinant adeno-associated virus vectors efficiently and persistently transduce chondrocytes in normal and osteoarthritic human articular cartilage. Hum Gene Ther 14:393–402
84. Hirschmann F, Verhoeyen E, Wirth D, Bauwens S, Hauser H, Rudert M (2002) Vital marking of articular chondrocytes by retroviral infection using green fluorescence protein. Osteoarthritis Cartilage 10:109–118
85. Shakibaei M, Seifarth C, John T, Rahmanzadeh M, Mobasheri A (2006) Igf-I extends the chondrogenic potential of human articular chondrocytes in vitro: molecular association between Sox9 and Erk1/2. Biochem Pharmacol 72:1382–1395
86. Sung L-Y, Chiu H-Y, Chen H-C, Chen Y-L, Chuang C-K, Hu Y-C (2009) Baculovirus-mediated growth factor expression in dedifferentiated chondrocytes accelerates redifferentiation: effects of combinational transduction. Tissue Eng Part A 15:1353–1362
87. Nesic D, Whiteside R, Brittberg M, Wendt D, Martin I, Mainil-Varlet P (2006) Cartilage tissue engineering for degenerative joint disease. Adv Drug Deliv Rev 58:300–322
88. Chen H-C, Sung L-Y, Lo W-H, Chuang C-K, Wang Y-H, Lin J-L et al (2008) Combination of baculovirus-mediated BMP-2 expression and rotating-shaft bioreactor culture synergistically enhances cartilage formation. Gene Ther 15:309–317
89. Saini S, Wick TM (2003) Concentric cylinder bioreactor for production of tissue engineered cartilage: effect of seeding density and hydrodynamic loading on construct development. Biotechnol Prog 19:510–521

90. Bueno EM, Bilgen B, Barabino GA (2005) Wavy-walled bioreactor supports increased cell proliferation and matrix deposition in engineered cartilage constructs. Tissue Eng 11:1699–1709
91. Carver SE, Heath CA (1999) Increasing extracellular matrix production in regenerating cartilage with intermittent physiological pressure. Biotechnol Bioeng 62:166–174
92. Freed LE, Langer R, Martin I, Pellis NR, VunjakNovakovic G (1997) Tissue engineering of cartilage in space. Proc Natl Acad Sci U S A 94:13885–13890
93. Hoch DH, Grodzinsky AJ, Koob TJ, Albert ML, Eyre DR (1983) Early changes in material properties of rabbit articular cartilage after meniscectomy. J Orthop Res 1:4–12
94. Madry H, Cucchiarini M, Stein U, Remberger K, Kohn D, Trippel SB (2003) Sustained transgene expression in cartilage defects in vivo after transplantation of articular chondrocytes modified by lipid-mediated gene transfer in a gel suspension delivery system. J Gene Med 5:502–509
95. Zachos TA, Diggs A, Weisbrode S, Bartlett J, Bertone AL (2007) Mesenchymal stem cell-mediated gene delivery of bone morphogenetic protein-2 in an articular fracture model. Mol Ther 15:1543–1550
96. Hanada K, Solchaga LA, Caplan AI, Hering TM, Goldberg VM, Yoo JU et al (2001) BMP-2 induction and TGF-β1 modulation of rat periosteal cell chondrogenesis. J Cell Biochem 81:284–294
97. Caplan AI, Bruder SP (2001) Mesenchymal stem cells: building blocks for molecular medicine in the 21st century. Trends Mol Med 7:259–264
98. Bruder SP, Jaiswal N, Haynesworth SE (1997) Growth kinetics, self-renewal, and the osteogenic potential of purified human mesenchymal stem cells during extensive subcultivation and following cryopreservation. J Cell Biochem 64:278–294
99. Carlberg AL, Pucci B, Rallapalli R, Tuan RS, Hall DJ (2001) Efficient chondrogenic differentiation of mesenchymal cells in micromass culture by retroviral gene transfer of BMP-2. Differentiation 67:128–138
100. Wakitani S, Goto T, Pineda SJ, Young RG, Mansour JM, Caplan AI et al (1994) Mesenchymal cell-based repair of large, full-thickness defects of articular cartilage. J Bone Joint Surg Am 76:579–592
101. Caplan AI, Elyaderani M, Mochizuki Y, Wakitani S, Goldberg VM (1997) Principles of cartilage repair and regeneration. Clin Orthop 342:254–269
102. Pelttari K, Winter A, Steck E, Goetzke K, Hennig T, Ochs BG et al (2006) Premature induction of hypertrophy during in vitro chondrogenesis of human mesenchymal stem cells correlates with calcification and vascular invasion after ectopic transplantation in SCID mice. Arthritis Rheum 54:3254–3266
103. Wang W, Li B, Li Y, Jiang Y, Ouyang H, Gao C (2010) In vivo restoration of full-thickness cartilage defects by poly(lactide-co-glycolide) sponges filled with fibrin gel, bone marrow mesenchymal stem cells and DNA complexes. Biomaterials 31:5953–5965
104. Marx JC, Allay JA, Persons DA, Nooner SA, Hargrove PW, Kelly PF et al (1999) High-efficiency transduction and long-term gene expression with a murine stem cell retroviral vector encoding the green fluorescent protein in human marrow stromal cells. Hum Gene Ther 10:1163–1173
105. Gazit D, Turgeman G, Kelley P, Wang E, Jalenak M, Zilberman Y et al (1999) Engineered pluripotent mesenchymal cells integrate and differentiate in regenerating bone: a novel cell-mediated gene therapy. J Gene Med 1:121–133
106. Martinek V, Fu FH, Lee CW, Huard J (2001) Treatment of osteochondral injuries – genetic engineering. Clin Sports Med 20:403–416, viii
107. Ho Y-C, Chung Y-C, Hwang S-M, Wang K-C, Hu Y-C (2005) Transgene expression and differentiation of baculovirus-transduced human mesenchymal stem cells. J Gene Med 7:860–868
108. Ho Y-C, Lee H-P, Hwang S-M, Lo W-H, Chen H-C, Chung C-K et al (2006) Baculovirus transduction of human mesenchymal stem cell-derived progenitor cells: variation of transgene expression with cellular differentiation states. Gene Ther 13:1471–1479

109. Palmer GD, Steinert A, Pascher A, Gouze E, Gouze JN, Betz O et al (2005) Gene-induced chondrogenesis of primary mesenchymal stem cells *in vitro*. Mol Ther 12:219–228
110. Steinert AF, Palmer GD, Pilapil C, Nöth U, Evans CH, Ghivizzani SC (2008) Enhanced in vitro chondrogenesis of primary mesenchymal stem cells by combined gene transfer. Tissue Eng Part A 15:1127–1139
111. Kim HJ, Im GI (2011) Electroporation-mediated transfer of SOX trio genes (SOX-5, SOX-6, and SOX-9) to enhance the chondrogenesis of mesenchymal stem cells. Stem Cells Dev 20:2103–2114
112. Park JS, Yang HN, Woo DG, Jeon SY, Do HJ, Lim HY et al (2011) Chondrogenesis of human mesenchymal stem cells mediated by the combination of SOX trio SOX5, 6, and 9 genes complexed with PEI-modified PLGA nanoparticles. Biomaterials 32:3679–3688
113. Mahmoudifar N, Doran PM (2010) Chondrogenic differentiation of human adipose-derived stem cells in polyglycolic acid mesh scaffolds under dynamic culture conditions. Biomaterials 31:3858–3867
114. Santo VE, Gomes ME, Mano JF, Reis RL (2013) Controlled release strategies for bone, cartilage, and osteochondral engineering-Part II: challenges on the evolution from single to multiple bioactive factor delivery. Tissue Eng Part B Rev 19:327–352
115. Freyria A-M, Mallein-Gerin F (2012) Chondrocytes or adult stem cells for cartilage repair: the indisputable role of growth factors. Injury 43:259–265
116. Diekman BO, Estes BT, Guilak F (2010) The effects of BMP6 overexpression on adipose stem cell chondrogenesis: interactions with dexamethasone and exogenous growth factors. J Biomed Mater Res A 93:994–1003
117. Im GI, Kim HJ, Lee JH (2011) Chondrogenesis of adipose stem cells in a porous PLGA scaffold impregnated with plasmid DNA containing SOX trio (SOX-5,-6 and -9) genes. Biomaterials 32:4385–4392
118. Sophia Fox AJ, Bedi A, Rodeo SA (2009) The basic science of articular cartilage: structure, composition, and function. Sports Health 1:461–468
119. Keeney M, Lai JH, Yang F (2011) Recent progress in cartilage tissue engineering. Curr Opin Biotechnol 25:734–740
120. Chen J, Chen H, Li P, Diao H, Zhu S, Dong L et al (2011) Simultaneous regeneration of articular cartilage and subchondral bone in vivo using MSCs induced by a spatially controlled gene delivery system in bilayered integrated scaffolds. Biomaterials 32:4793–4805
121. Yang HN, Park JS, Woo DG, Jeon SY, Do HJ, Lim HY et al (2011) Chondrogenesis of mesenchymal stem cells and dedifferentiated chondrocytes by transfection with SOX Trio genes. Biomaterials 32:7695–7704

# Chapter 5
# Conclusions and Perspectives

**Abstract** Gene therapy has been widely explored for the treatment of various diseases and disorders, yet successful applications of gene therapy in bone and cartilage engineering are still not in the near horizon. This chapter summarizes the major hurdles that set the roadblocks to the application of gene therapy in tissue engineering and discusses future perspectives.

## 5.1 Concluding Remarks on Gene Therapy in Bone Tissue Engineering

Although gene therapy has shown great promise in clinical trials for the treatment of arthritis [1, 2] and other genetic diseases [3], the safety issues continue to impede the translation of gene therapy-based bone healing/regeneration from bench to bedside. This is particularly true because bone fractures/defects are typically non-lethal, which make the patients and physicians reluctant to adopt the gene transfer approach. To ease the safety concern, systematic assessment of the potential immune responses and genotoxicity resulting from the vector or transgene products is imperative.

### *5.1.1 Immune Responses Against Viral Vectors/Transgenes*

Immune responses elicited by the gene delivery vectors have been a major roadblock to the gene delivery-based bone engineering. Even non-viral vectors in the form of DNA or RNA can activate toll-like receptors (TLR) such as TLR-3 and

Y.-C. Hu, *Gene Therapy for Cartilage and Bone Tissue Engineering*, SpringerBriefs in Bioengineering, DOI 10.1007/978-3-642-53923-7_5,

TLR-9 [4]. It is unraveled that direct injection of adenovirus that expresses bone morphogenetic protein 2 (BMP-2) into sheeps elicits anti-BMP-2 and anti-adenovirus immune responses, which is correlated with the failure of bone healing [5]. Nonetheless, bone healing is evident in horses that receive direct adenovirus injection, although anti-adenovirus antibodies is detectable [6]. Conversely, implantation of bone marrow-derived mesenchymal stem cells (BMSCs) transduced with the adenovirus expressing BMP-2 into segmental defects in goats triggers temporary cellular and persistent humoral immune responses against adenovirus [7]. Such immune responses may eliminate the transduced BMSCs, shorten the duration of BMP-2 expression and impair the effectiveness of bone healing. Nonetheless, successful repair of goat tibial bone defects is observed, suggesting that the immune response elicited as a result of ex vivo therapy may not be strong enough to hinder successful bone regeneration.

In favor of this notion, baculovirus transduction of BMSCs elicits transient and mild innate response in vitro [8] and implantation of baculovirus-engineered BMSCs into animals triggers transient immune responses [9]. Notably, baculovirus transduction of human BMSCs perturbs the expression of 816 genes and activates the TLR-3 pathway, leading to secretion of interleukin (IL)-6 and IL-8 [8]. Nonetheless, baculovirus-engineered BMSCs implanted into segmental bone defects result in successful healing [10]. Likewise, implantation of baculovirus-engineered ASCs that express BMP-2 and vascular endothelial growth factor (VEGF) heal the segmental bone defects in New Zealand white rabbits [11], although the implantation slightly elicits humoral and cellular immune responses against the transgene products [12].

### *5.1.2 Roles of Host Immunity on Bone Healing*

It is recently shown that bone fracture healing may be retarded by endogenous adaptive/innate immune responses. For instance, $\gamma/\delta$ T cells, the innate lymphocytes involved in tissue repair, can repress bone healing by influencing the fate of other responder cells and the ultimate callus formation [13]. Pro-inflammatory T cells also inhibit the ability of exogenously added BMSCs to mediate bone repair, owing to interferon (IFN)-$\gamma$–induced down-regulation of RunX2 pathway and enhancement of tumor necrosis factor (TNF)-$\alpha$ signaling in the stem cells [14]. Conversely, reduction of IFN-$\gamma$ and TNF-$\alpha$ concentrations, by systemic infusion of Foxp3$^+$ regulatory T cells or by local administration of aspirin, markedly improves BMSCs-based calvarial defect repair in mice [14]. Furthermore, delayed fracture healing correlates with enhanced levels of terminally differentiated CD8$^+$ effector memory T (TEMRA) cells in peripheral blood [15]. These CD8$^+$ TEMRA cells are enriched in fracture hematoma and are the major producers of IFN-$\gamma$/TNF-$\alpha$, which inhibit osteogenic differentiation [15]. These data collectively underscore the crucial role of recipient T cells in BMSCs-based bone engineering [14].

### 5.1.3 *Genotoxicity*

Another critical issue relevant to gene therapy is genotoxicity, especially for viral vectors [16]. Although a wealth of literature has reported the use of viral vectors for stem cell transduction and bone regeneration, whether the genetic modification provokes aberrant host gene expression, tumorigenesis or heterotopic ossification is rarely assessed. In this regard, the safety of baculovirus for stem cell engineering and bone regeneration has been evaluated. We uncover that baculovirus transduction does not impair the ability of BMSCs to differentiate towards different lineages [17]. In vitro transduction of human BMSCs with the FLP/Frt-based hybrid baculoviral vectors (see Chap. 2) neither integrates the transgene into the host chromosome nor disrupts the karyotype of BMSCs [18]. Neither do the transduced human BMSCs induce tumor formation after implantation into nude mice, thus supporting the safety of baculovirus-transduced BMSCs for cell therapy [18]. Implantation of the FLP/Frt-based hybrid baculovirus-engineered adipose-derived stem cells (ASCs) that persistently express BMP-2/VEGF heal and remodel the massive segmental defects in New Zealand White rabbits. Very importantly, the baculovirus-engineered cells are eradicated after 4 weeks of implantation [12]. The clearance of virus-transduced cells may imply that the concurrent removal of viral vector-associated nucleic acids and thus minimize the potential side effects. Furthermore, X-ray radiography demonstrates no heterotopic bone formation while positron emission tomography/computed tomography (PET/CT) scans reveal no signs of tumor formation at 8 months after transplantation into rabbits (unpublished data). The eradication of transplanted cells, regardless of being genetically engineered, has been reported in numerous studies, thus suggesting the safe use of ex vivo gene therapy for bone regeneration.

## 5.2 Concluding Remarks on Gene Therapy in Cartilage Tissue Engineering

Gene therapy in combination with tissue engineering offers a promising solution for cartilage repair. However, several hurdles remain to be solved. First, chondroinduction of BMSCs with growth factors is often accompanied by osteogenesis and hypertrophy, which could lead to apoptosis and calcification [19–22]. Although ASCs may be another promising cell source, ASCs are inferior to BMSCs [23] and chondrocytes [24] in terms of chondrogenesis potential. Moreover, chondroinduction of ASCs (e.g. with transforming growth factor β3 (TGF-β3) and BMP-6) is still associated with hypertrophy in vitro and calcification in vivo [25]. In agreement with the findings, short-term baculovirus-mediated expression of TGF-β3/BMP-6 in ASCs promotes chondrogenesis, but ASCs also undergo osteogenesis and hypertrophy in vitro and the cell/scaffold constructs lead to signs of degeneration and ossification at 24 weeks post-implantation into full-thickness cartilage defects in rabbits [26]. Nonetheless, the ASCs engineered with hybrid baculovirus vectors for sustained expression of TGF-β3/BMP-6 effectively promotes the chondrogenesis of ASCs while suppresses the osteogenesis and hypertrophy, enabling the constructs to form more mature cartilage-like tissues with improved cell morphology, extracellular matrix (ECM) deposition and mechanical properties in 2 weeks. After transplantation, the engineered cartilaginous constructs fill the critical-size defects at week 8 and develop into cartilages with an architecture that is characteristic of mature hyaline cartilages, in terms of matrix composition, organization and cell morphology. Strikingly, the neocartilages display a zonal structure characteristic of the native cartilage [26]. As such, the growth factor combination as well as expression level and duration are critical to guide the stem cell chondrogenesis while suppressing undesired hypertrophy/fibrocartilage formation.

Second, integration of regenerated cartilage with adjacent native cartilage is crucial for immediate functionality and long-term performance, because integration provides stable biologic fixation, load distribution, and also the proper mechanotransduction necessary for homeostasis. However, cartilage's hyaline, nonadhesive nature precludes integration and lateral integration of cartilage to adjacent cartilage is rarely reported, which presents a major stumbling block to the success and commercialization of cartilage engineering [21, 27]. Very commonly the cells genetically modified ex vivo are seeded to scaffold and implanted immediately implanted in vivo [28]. However, in vivo mechanical load to the implanted cell/scaffold constructs devoid of cartilage ECM may result in tissue deformation that impairs the regeneration process due to the lack of mechanical and weight-bearing support [29]. Additionally, the cells transplanted into full-thickness defects may not remain for long time periods [30] In contrast to repair approaches based on formation and maturation of new tissue in situ, it would be appealing to achieve the cartilaginous tissue regeneration by implanting a preformed graft [31]. However, to date it remains to be determined how closely the engineered grafts need to resemble the native tissues and which properties are more important than others [32]. Obradovic et al. have proposed that immature cartilaginous tissues have poorer mechanical properties but achieve better repair and integration than mature constructs [29]. However, it should be noted that upon loading, mismatches between the biomechanical properties of the cartilage implant and native tissue result in stress concentrations diminishing integration and damaging surrounding tissue. In accord, cartilaginous constructs derived from baculovirus-transduced chondrocytes [33] and ASCs [26] suggest that more mature constructs achieve better in vivo cartilage repair when compared with freshly seeded constructs. At this time, it remains debatable regarding 'how much is enough' and whether engineered cartilage needs to mimic the mechanical properties of native cartilage at the time of implantation [32]. Future studies to elucidate how mature the constructs are needed.

Third, restoration of articular cartilage-specific zonal structures is important but difficult to achieve. Failure to overcome these problems often leads to the regeneration of fibrocartilages with inferior mechanical properties, which ultimately collapse in the long term [21, 27]. Future studies should be directed towards enhancing the restoration of cartilage zonal structures [27] and prevention of angiogenesis/ossification [34].

## 5.3 Future Perspectives

To date, the clinical application of gene therapy in conjunction with tissue engineering is still not in the near horizon. One exciting advancement is that adeno-associated virus (AAV) has been the first vector approved for human gene therapy in European Union. However, relatively few studies have employed AAV for bone/cartilage engineering, which may partly stem from the difficulty and high cost associated with the production of AAV vectors (which requires transfection of producer cells with multiple plasmids). The new AAV production method using the baculovirus/insect cell system [35] and approval of Glybera®, an AAV vector produced using the baculovirus/insect cell system, for gene therapy may encourage wider applications of AAV vectors in bone/cartilage tissue engineering. In addition, baculovirus, an emerging gene delivery vector based on the non-pathogenic insect virus, holds promise for bone and cartilage regeneration, especially multiple preclinical studies have demonstrated the safety and efficacy of baculovirus-engineered stem cells in bone and cartilage regeneration models. Other new vectors with improved safety features (such as integrase-deficient lentiviral vectors) may be worth of exploring.

## References


1. Evans CH, Ghivizzani SC, Robbins PD (2009) Progress and prospects: genetic treatments for disorders of bones and joints. Gene Ther 16:944–952
2. Evans CH, Ghivizzani SC, Robbins PD (2009) Orthopedic gene therapy in 2008. Mol Ther 17:231–244
3. Seymour LW, Thrasher AJ (2012) Gene therapy matures in the clinic. Nat Biotechnol 30:588–593
4. Takeuchi O, Akira S (2010) Pattern recognition receptors and inflammation. Cell 140: 805–820
5. Egermann M, Lill CA, Griesbeck K, Evans CH, Robbins PD, Schneider E et al (2006) Effect of BMP-2 gene transfer on bone healing in sheep. Gene Ther 13:1290–1299
6. Ishihara A, Shields KM, Litsky AS, Mattoon JS, Weisbrode SE, Bartlett JS et al (2008) Osteogenic gene regulation and relative acceleration of healing by adenoviral-mediated transfer of human BMP-2 or -6 in equine osteotomy and ostectomy models. J Orthop Res 26:764–771
7. Xu XL, Tang T, Dai K, Zhu Z, Guo XE, Yu C et al (2005) Immune response and effect of adenovirus-mediated human BMP-2 gene transfer on the repair of segmental tibial bone defects in goats. Acta Orthop 76:637–646
8. Chen G-Y, Shiah H-C, Su H-J, Chen C-Y, Chuang Y-J, Lo W-H et al (2009) Baculovirus transduction of mesenchymal stem cells triggers the toll-like receptor 3 (TLR3) pathway. J Virol 83:10548–10556
9. Chuang C-K, Wong T-H, Hwang S-M, Chang Y-H, Chen Y-H, Chiu Y-C et al (2009) Baculovirus transduction of mesenchymal stem cells: in vitro responses and in vivo immune responses after cell transplantation. Mol Ther 17:889–896
10. Lin C-Y, Chang Y-H, Lin K-J, Yen T-Z, Tai C-L, Chen C-Y et al (2010) The healing of critical-sized femoral segmental bone defects in rabbits using baculovirus-engineered mesenchymal stem cells. Biomaterials 31:3222–3230
11. Lin C-Y, Lin K-J, Kao C-Y, Chen M-C, Yen T-Z, Lo W-H et al (2011) The role of adipose-derived stem cells engineered with the persistently expressing hybrid baculovirus in the healing of massive bone defects. Biomaterials 32:6505–6514

12. Lin C-Y, Lin K-J, Li K-C, Sung L-Y, Hsueh S, Lu C-H et al (2012) Immune responses during healing of massive segmental femoral bone defects mediated by hybrid baculovirus-engineered ASCs. Biomaterials 33:7422–7434
13. Colburn NT, Zaal KJM, Wang F, Tuan RS (2009) A role for γ/δ T cells in a mouse model of fracture healing. Arthritis Rheum 60:1694–1703
14. Liu Y, Wang L, Kikuiri T, Akiyama K, Chen C, Xu X et al (2012) Mesenchymal stem cell-based tissue regeneration is governed by recipient T lymphocytes via IFN-γ and TNF-α. Nat Med 17:1594–1601
15. Reinke S, Geissler S, Taylor WR, Schmidt-Bleek K, Juelke K, Schwachmeyer V et al (2013) Terminally differentiated CD8+ T cells negatively affect bone regeneration in humans. Sci Transl Med 5:177ra36
16. Huang S, Kamihira M (2013) Development of hybrid viral vectors for gene therapy. Biotechnol Adv 31:208–223
17. Ho Y-C, Lee H-P, Hwang S-M, Lo W-H, Chen H-C, Chung C-K et al (2006) Baculovirus transduction of human mesenchymal stem cell-derived progenitor cells: variation of transgene expression with cellular differentiation states. Gene Ther 13:1471–1479
18. Chen C-Y, Wu H-H, Chen C-P, Chern S-R, Hwang S-M, Huang S-F et al (2011) Biosafety assessment of human mesenchymal stem cells engineered by hybrid baculovirus vectors. Mol Pharm 8:1505–1514
19. Pelttari K, Winter A, Steck E, Goetzke K, Hennig T, Ochs BG et al (2006) Premature induction of hypertrophy during in vitro chondrogenesis of human mesenchymal stem cells correlates with calcification and vascular invasion after ectopic transplantation in SCID mice. Arthritis Rheum 54:3254–3266
20. Mueller MB, Tuan RS (2008) Functional characterization of hypertrophy in chondrogenesis of human mesenchymal stem cells. Arthritis Rheum 58:1377–1388
21. Huey DJ, Hu JC, Athanasiou KA (2012) Unlike bone, cartilage regeneration remains elusive. Science 338:917–921
22. Mahmoudifar N, Doran PM (2012) Chondrogenesis and cartilage tissue engineering: the longer road to technology development. Trends Biotechnol 30:166–176
23. Afizah H, Yang Z, Hui JH, Ouyang HW, Lee EH (2007) A comparison between the chondrogenic potential of human bone marrow stem cells (BMSCs) and adipose-derived stem cells (ADSCs) taken from the same donors. Tissue Eng 13:659–666
24. Mahmoudifar N, Doran PM (2010) Extent of cell differentiation and capacity for cartilage synthesis in human adult adipose-derived stem cells: comparison with fetal chondrocytes. Biotechnol Bioeng 107:393–401
25. Hennig T, Lorenz H, Thiel A, Goetzke K, Dickhut A, Geiger F et al (2007) Reduced chondrogenic potential of adipose tissue derived stromal cells correlates with an altered TGFbeta receptor and BMP profile and is overcome by BMP-6. J Cell Physiol 211:682–691
26. Lu C-H, Yeh T-S, Yeh C-L, Fang Y-HD, Sung L-Y, Lin S-Y et al (2013) Regenerating cartilages by engineered ASCs: prolonged TGF-β3/BMP-6 expression improved articular cartilage formation and restored zonal structure. Mol Ther 22:186–195
27. Keeney M, Lai JH, Yang F (2011) Recent progress in cartilage tissue engineering. Curr Opin Biotechnol 25:734–740
28. Madry H, Kaul G, Cucchiarini M, Stein U, Zurakowski D, Remberger K et al (2005) Enhanced repair of articular cartilage defects *in vivo* by transplanted chondrocytes overexpressing insulin-like growth factor I (IGF-I). Gene Ther 12:1171–1179
29. Obradovic B, Martin I, Padera RF, Treppo S, Freed LE, Vunjak-Novakovic G (2001) Integration of engineered cartilage. J Orthop Res 19:1089–1097
30. Mierisch CM, Wilson HA, Turner MA, Milbrandt TA, Berthoux L, Hammarskjold ML et al (2003) Chondrocyte transplantation into articular cartilage defects with use of calcium alginate: the fate of the cells. J Bone Joint Surg Am 85A:1757–1767
31. Gelse K, Schneider H (2006) *Ex vivo* gene therapy approaches to cartilage repair. Adv Drug Deliv Rev 58:259–284

32. Grayson WL, Chao P-HG, Marolt D, Kaplan DL, Vunjak-Novakovic G (2008) Engineering custom-designed osteochondral tissue grafts. Trends Biotechnol 26:181–189
33. Chen H-C, Chang Y-H, Chuang C-K, Lin C-Y, Sung L-Y, Wang Y-H et al (2009) The repair of osteochondral defects using baculovirus-mediated gene transfer with de-differentiated chondrocytes in bioreactor culture. Biomaterials 30:674–681
34. Sheyn D, Mizrahi O, Benjamin S, Gazit Z, Pelled G, Gazit D (2010) Genetically modified cells in regenerative medicine and tissue engineering. Adv Drug Deliv Rev 62:683–698
35. Smith RH, Levy JR, Kotin RM (2009) A simplified baculovirus-AAV expression vector system coupled with one-step affinity purification yields high-titer rAAV stocks from insect cells. Mol Ther 17:1888–1896

Printed by Publishers' Graphics LLC